AF291375

Road Engineering

Hui Li · Fangyu Liu · Jingbin Yang · Xue Zhang ·
Ming Jia

Road Engineering

Hui Li
College of Transportation
Tongji University
Shanghai, China

Fangyu Liu
College of Civil Engineering
Tongji University
Shanghai, China

Jingbin Yang
College of Transportation
Tongji University
Shanghai, China

Xue Zhang
College of Transportation
Shandong University of Science
and Technology
Qingdao, Shandong Province, China

Ming Jia
College of Transportation
Shandong University of Science
and Technology
Qingdao, Shandong Province, China

ISBN 978-981-95-6658-7   ISBN 978-981-95-6659-4   (eBook)
https://doi.org/10.1007/978-981-95-6659-4

Jointly published with Tongji University Press Co., Ltd.
The print edition is not for sale in China (Mainland). Customers from China (Mainland) please order the print book from: Tongji University Press Co., Ltd.

This Springer imprint is published by the registered company Springer Nature Singapore Pte Ltd.
The registered company address is: 152 Beach Road, #21-01/04 Gateway East, Singapore 189721, Singapore

If disposing of this product, please recycle the paper.

Acknowledgments

I would also like to express my profound gratitude to Zhenhua Dai, Yang Sun, Ziyi Tang, Yuzhao Han, Kuigeng Wang, Jie Pan, Lei Wang, Saad Khan, Jiaxing Ren, Jie Yang, Bing Yang, Zhen Dai, and Jiawen Liu for their valuable guidance and generous support, which greatly enhanced the quality of this work. This work was supported by the Graduate Textbook Development Project of Tongji University (Grant No. 2022JC41).

Contents

About the Authors

Dr. Hui Li is a professor in the College of Transportation at Tongji University, Shanghai, China and a research scientist in the Department of Civil and Environmental Engineering at the University of California, Davis. He received his Ph.D. from the University of California, Davis, and his master's degree from Southeast University. He is a registered civil engineer in California, a member of the American Transportation and Environment Committee, and a member of the Youth Expert Committee of the China Highway Society. Dr. Li's research focuses on the sustainability, resilience and intelligence of urban transportation infrastructure. He has published more than 60 papers, authored 3 monographs, obtained over 30 patents, contributed to five technical standards, and received five provincial or ministerial-level awards. He serves as a Young Communication Expert for the Engineering Journal of the Chinese Academy of Engineering, deputy editor-in-chief and editorial board member of IJTST, TRD and other journals, and is active on several prominent international academic committees. His research outcomes have been applied to the construction of national sponge cities and green smart road projects in China, including Xiong' an New Area, the Beijing Winter Olympic Games, the Hangzhou Asian Games, and the Qinghai Tibet Highway.

Dr. Fangyu Liu is a research professor in the College of Civil Engineering at Tongji University, Shanghai, China. He previously conducted postdoctoral research at the University of Illinois Urbana-Champaign. He earned his Ph.D. from Virginia Tech and received both his master's and bachelor's degrees from Tongji University. His research focuses primarily on the AI-driven smart and resilient transportation infrastructures. He has published more than 30 papers.

Dr. Jingbin Yang has long been engaged in the research on high-performance pavement material strengthening and toughening design, intelligent directional control of admixtures, and durability design of eco-friendly low-carbon pavement materials. A series of theories and methods for improving the service life of infrastructure have been constructed, aiming to overcome scientific problems such as the moisture migration behavior and accurate performance regulation of cement-based materials.

He has presided over 1 youth project of the National Natural Science Foundation of China, 1 general project of the China Postdoctoral Science Foundation, and 1 open fund project of the Key Laboratory of Water Engineering Materials of the Ministry of Water Resources. He serves as the Secretary-General of the Solid Waste Eco-material Admixtures Academic Committee of the Solid Waste Branch of the Chinese Silicate Society.

Dr. Xue Zhang is a lecturer (university-appointed associate professor based on University Talent Recruitment Program) at the College of Transportation, Shandong University of Science and Technology. She received her PhD in Transportation Engineering from Tongji University through a direct doctoral program. Her research focuses on sustainable road transportation infrastructure, ecological and functional pavement materials, heat-reflective cooling pavement technology, and low-carbon utilization of solid waste. She has participated in three national research projects, including the National Key R&D Program of China, and three provincial or ministerial projects. She has also been involved in several engineering projects, including the Taizicheng Service Area of the Yanqing-Chongli Expressway for the Beijing Winter Olympics. She has published 17 SCI/EI-indexed papers, with the highest impact factor of 12.0, and holds 10 authorized invention patents, including two US and UK patents. She has contributed to one group technical standard of the China Highway and Transportation Society and delivered six presentations at international conferences, including TRB. She serves as a reviewer for several SCI journals and as an English guest editor of Concrete and Cement Products.

Dr. Ming Jia is a postdoctoral teaching fellow in the College of Transportation at Shandong University of Science and Technology. He received his Ph.D. from the College of Transportation Engineering at Tongji University. Dr. Jia has long been engaged in research on sustainable and resilient transportation infrastructure. His research interests mainly include the control of asphalt fumes, road surface runoff pollution, and light–thermal pollution in roadway environments. He is currently leading two research projects and has participated in three National Key R&D Program projects of China. He has published more than 20 papers in related fields, obtained over 10 invention patents in the United States, the United Kingdom, and China, and contributed to three technical standards. He also serves as a reviewer for over ten international journals, including Journal of Hazardous Materials, Chemical Engineering Journal, Journal of Cleaner Production, Construction and Building Materials and so on.

Chapter 1
Introduction

1.1 Development of Road Engineering

The development of roads is inseparable from the development of human beings. Primitive people walked in the mountains and rivers, and the tracks formed roads over the years. After humans began to settle, roads spread out around settlements to faciliate commerce. Human social activities created roads, and the creation and development of roads contributed to the progress of human beings. Road engineering has gradually evolved from experience-based construction into a multidisciplinary field integrating materials science, structural mechanics, geotechnical engineering, traffic engineering, environmental science, and digital technologies[4, 5].

1.1.1 Development of Western Roads

Before 1900 B.C., the Assyrian Empire built roads radiating from Babylon, and some of these roads still exist today between Baghdad and Isfahan. Historical records suggest that the ancient African nation of Carthage (600 B.C.–14 B.C.) was among the earliest to construct paved roads, which Rome later used. More broadly, early road systems in the ancient Near East and Mediterranean world were closely associated with military movement, administration, and long-distance trade, which created the social and economic conditions for the emergence of more permanent road infrastructure[6].

The grand construction of roads in the Roman Empire played a significant role in maintaining the empire's prosperity. By building roads from the capital of Rome to Italy, Britain, France, Spain, Germany, parts of Asia Minor, Arabia and northern Africa, the empire maintained dominance in these regions. The network included 322 major routes with a total length of 78,000 km (52,964 Roman miles). The Roman road network was dominated by 29 main roads, the most famous of which is the

© Tongji University Press Co., Ltd. 2026

H. Li et al., *Road Engineering*, https://doi.org/10.1007/978-981-95-6659-4_1

Via Appius, which runs from southeastern Rome over the Apennine Mountains to Brindisi, with a total length of approximately 660 km. This network began to be built at approximately 400 B.C. and was constructed over several decades, serving as a link between Rome and regions within the Roman Empire. Roman roads were often connected in straight lines despite the rugged terrain, and traces of tunnels, bridges and retaining walls remain today. Several of the main military roads were 11–12 m wide, with the middle part being 3.7–4.9 m wide and paved with hard materials for infantry use. Causeways approximately 0.6 m wide were built on both sides of the road, probably for officers' commands, with cavalry paths 2.4 m wide on each side. The construction method involved excavating the road trench first and then building it in four layers with different sizes of stones and mud or mortar, with a total thickness of about 1 m. The style of the road surface also varied. The more advanced Via Appia used uneven stone slabs of 1–1.5 m in length, which were brought from 160 km away and set in mortar. With the fall of the Roman Empire, road construction also declined. The rise and fall of states and the condition of the roads are closely related. According to Encyclopaedia Britannica, the Roman system extended from Britain to the Tigris–Euphrates river system and from the Danube to Spain and northern Africa, and the Romans built about 80,000 km of hard-surfaced highways, primarily for military purposes. This historical experience shows that route continuity, structural layering, and drainage were already recognized as key elements affecting road durability and serviceability[4, 5].

In Western countries, the first person to adopt scientific methods to improve road construction was the French engineer Trésaguet in the Napoleonic era, whose work marked the first step toward scientific and modern road-building techniques and contributed to the development of the French road network during Napoleon's reign (1804–1814), thus being honored as the father of modern road construction. Telford, a Scottish engineer in England, built a road in 1815 using a large-stone base pavement structure. In 1816, another Scottish engineer in England, Macadam, conducted a study of the damage mechanism of gravel pavement and advocated replacing the bulky large stone base introduced by Telford with small-sized gravel, which became known as Macadam pavement [1]. After the invention of the stone rolling machine in 1858, the development of gravel pavement was promoted. Later, the horse-drawn roller was used to compact the pavement. In 1860, the emergence of the steam roller in France further advanced pavement construction technology. In 1888, Telford served as the first president of the Civil Engineers. In the early twentieth century, gravel pavement was widely recognized as an effective pavement type. In 1919, one of the earliest freeway, the AVUS, was opened in Germany [2].

Since 1945, road development in many countries has entered the stage of modern roads systems. European countries, the United States and Japan have built relatively complete national road networks. Road transportation has played a leading role in comprehensive transport systems. Today, road construction is becoming more intelligent, environmentally friendly and sustainable. A further landmark in the scientific development of pavement engineering was the AASHO Road Test (1958–1960), which established fundamental relationships among axle load, pavement performance, and structural design, and profoundly influenced modern pavement design methods used for both asphalt and concrete pavements[11]. In recent decades, the focus of Western road development has further expanded from providing traffic capacity to improving durability, life-cycle performance, resource efficiency, carbon reduction, and resilience to climate change. Recent review studies have emphasized that pavement engineering is increasingly evaluated from a systems perspective involving sustainability assessment, recycling technologies, circular-economy strategies, and adaptation to temperature extremes, precipitation change, and sea-level rise[7–10].

1.1.2 Development of Chinese Roads

Ancient Chinese roads once led the world and left an important chapter in the history of transportation. According to the Records of the Grand Historian, carts and roads for travel already existed in China more than 4,000 years ago. During the Shang Dynasty (1600 B.C.–1046 B.C.), stagecoach transport began. The Western Zhou Dynasty (1046 B.C.–771 B.C.) created a road system centered on the capital and a relatively sound road management system. The Qin Dynasty (221 B.C.–206 B.C.) established a large-scale road transport network with a total mileage of more than 12,000 km by constructing the Coach Road and Straight Road. During the Western Han Dynasty (206 B.C.–25 A.D.), about 30,000 stagecoach stations were established, and road transportation became more developed, especially with the opening of the Silk Road connecting Eurasia, which contributed to economic and cultural exchanges between Eastern and Western countries. During the Tang Dynasty (618 A.D.–907 A.D.), a period of economic and cultural prosperity in ancient China, a network of approximately 22,000 km of post roads was built, centering on Chang'an city (now Xi'an). During the Song, Yuan, Ming and Qing Dynasties (960 A.D.–1911 A.D.), road transportation further developed. Modern reference works on Chinese road history also note that traces of road pavement have been found at Shang Dynasty sites, that the Qin state built an extensive radial network after unification, and that successive dynasties further improved imperial roads, courier routes, and transport management. These facts indicate that ancient Chinese road development was characterized not only by large scale, but also by administrative organization and engineering continuity over long historical periods[12].

Although China once created an ancient road culture, the development of road construction in modern China was relatively slow due to long-term feudal rule and external conflicts in the last century. At the end of the Qing Dynasty, some rudimentary roads built on the original post roads began to appear. Roads initially developed during the Republic of China period (1912–1949), and about 130,000 km were constructed. Most of these roads were substandard, with rudimentary facilities and poor road conditions. By 1949, only 80,000 km remained open to traffic, one-third of the counties had no roads, and there were no roads in Tibet. Automobile transportation began with the first automobile imported from abroad in 1901, and by 1949, there were approximately 50,000 automobiles, most of which were outdated.

Around the country, transportation still relied mainly on human and animal power. In other words, modern road construction in China before 1949 remained limited in both network coverage and technical standard, and it had not yet formed a modern highway system capable of supporting nationwide industrialization and large-scale motorized transport. This historical background helps explain why highway development after 1949 became a fundamental component of national reconstruction and economic modernization[13].

Since the founding of the People's Republic of China in 1949, China has entered a period of rapid development. Owing to the rapid development of industrial and agricultural production and the gradual improvement in people's living standards, especially the establishment and development of the automobile and petroleum industries, China's highway transportation has developed rapidly. Especially after 1978, the country implemented a policy focusing on economic development, and road construction also entered a new stage. By the end of 2018, the total road mileage had reached 4,846,500 km, and road passenger and freight traffic reached 13,672 million passengers and 39,569 million tons, respectively. Road transportation has penetrated all aspects of economic and social life, occupying an increasingly important position in the national economy.

In the mid-1980s, China began to build expressways, and over the past 30 years, initial expressway corridors connecting important cities and regions have formed. Many economically developed regions are forming expressway trunk networks. The construction and use of expressways provide suitable conditions for the fast, efficient, safe and comfortable operation of automobiles, indicating that China's road transportation has entered a new era [3]. The total length and density of roads increase each year, and the proportion of maintenance also increases each year. Furthermore, approximately 10% of China's expressways require major repairs each year. Intelligent networked road infrastructure and roadside infrastructure have become new trends in China's road construction. Recent international research further suggests that road infrastructure development is now increasingly shaped by three interrelated directions: smart-road technologies, sustainable maintenance, and climate-resilient planning. Therefore, the current development of Chinese roads should be understood not only as an expansion of expressway mileage, but also as a transition toward higher-quality infrastructure characterized by intelligence, sustainability, resilience, and whole-life asset management[14–16].

1.2 Classification of Pavement

Pavements are engineered structures constructed on the earth's surface to facilitate the movement of people and goods. Roads can be classified into highways, urban roads and special roads based on their usage characteristics.

1.2.1 Road Types

Highways

Highways are roads designed and equipped to specified technical standards to connect cities, towns, and other settlements, primarily for motor-vehicle traffic. Owing to their administrative status and network significance, highways can be divided into national highways, provincial highways, prefectural highways and country highways.

(1) National highways are part of the national trunk road network, with national political, economic and defense significance, and are designated as national trunk highways.

(2) Provincial highways are part of the provincial highway network, with provincial political, economic and defense significance, and are designated as provincial trunk highways.

(3) Prefectural highways are highways of prefectural political and economic significance within the regional road network.

(4) Country highways are roads built in rural areas that mainly serve local transport and connect county seats, townships, villages, and other local nodes.

Urban Roads

Urban roads are roads within urban areas that are designed and equipped to specified technical standards for vehicular and pedestrian traffic.(1) Expressways / Urban Expressways: High-speed roads for long-distance urban traffic, with limited access and few intersections. (2) Arterial Roads / Major Roads: Main urban roads that connect different districts and carry large traffic volumes. (3) Collector Roads: Roads that collect traffic from local streets and connect it to arterial roads. (4) Local Streets / Access Roads: Roads mainly serving residential areas, buildings, shops, and local access. (5) Service Roads / Frontage Roads: Roads running parallel to major roads, used for access to adjacent properties. 6. Pedestrian Streets Streets mainly designed for pedestrians, often located in commercial or historical areas.

Special Roads

Special roads, such as factory roads, mine roads and forest roads, are invested in and built by industrial, mining, agricultural and forestry departments[17, 18].

(1) Factory and Mine Roads

Factory and mine roads are roads dedicated to vehicles used for industrial and mining transport. They are usually divided into in-plant roads, out-of-plant roads and open-pit mine haul roads. Out-of-plant roads connect mining enterprises with national highways, urban roads, stations, ports or scattered workshops and residential areas.

(2) Forest Roads

Forest roads are built in forest areas, mainly for forestry transport. Owing to forest terrain and operational conditions, their technical requirements should comply with relevant technical standards for forest road engineering.

Because of variations in road locations, traffic properties and functions, the design basis, standards and specific requirements also differ. Therefore, the design and construction of pavements must be carried out in accordance with the corresponding technical specifications.

1.2.2 Classification of Pavement

Pavement types can be classified from different perspectives and are generally distinguished by the material used for the surface layer, such as cement concrete pavement, asphalt pavement and gravel pavement. However, in engineering design, pavements can be divided into three categories, namely asphalt concrete pavement, composite pavement and cement concrete pavement (also called rigid pavement), mainly because of the similarity of their mechanical properties. According to the types and combinations of base materials, asphalt concrete pavement can also be divided into flexible base asphalt pavement, semi-rigid base asphalt pavement, combined base asphalt pavement and rigid base asphalt pavement. Generally, cement concrete pavement and asphalt concrete pavement are referred to as paved pavements; surface treatment, asphalt gravel and asphalt penetration pavements are referred to as simple pavements; gravel pavement and other types are classified as unpaved pavements. Sand and gravel pavement is composed of sand and stone as aggregates, with soil, water, and ash as binding materials, formed according to a specific proportion, including graded sand (gravel) pavement, mud gravel pavement, water-bound gravel pavement, gap-graded gravel pavement and other gravel pavements.

Asphalt Bonded Base, Granular Base (Flexible Base) Asphalt Pavement

The overall structural stiffness of flexible base asphalt pavement is relatively low, and the surface deformation under vehicle loading is greater than that of semi-rigid base asphalt pavement. Although the tensile strength of each layer of the pavement structure is low, a reasonable structural combination and thickness design can ensure that the pavement structure has adequate resistance to load effects. At the same time, vehicle loads are transferred to the roadbed through each structural layer, which can control the compressive stress on the roadbed within a certain range. This type of pavement structure mainly relies on compressive and shear strength to withstand the effects of vehicle loads. Flexible base asphalt pavement mainly includes pavement structures composed of various untreated granular bases and different types of asphalt layers.

Inorganic Bonding Material Base (Semi-rigid Base) Asphalt Pavement

The base constructed with soil or gravel and industrial waste containing hydraulic bonding material, and treated with cement, lime or other inorganic bonding materials, is called the inorganic bonding material base, which has the mechanical properties of a flexible base in the early stage, whereas strength and stiffness increase substantially at later stages, although the final strength and stiffness are still lower than those of cement concrete. Because the stiffness of this material is between that of a flexible base and a rigid base, this base and the asphalt surface layer laid on it are also referred to as semi-rigid base asphalt pavement.

Combined Base Layer Asphalt Pavement

The base layer of asphalt pavement contains inorganically stabilized materials, cement concrete and other relatively stiff materials. However, flexible materials can exist between the asphalt layer and these stiff materials, such as pavement structures consisting of asphalt concrete + graded gravel + inorganically stabilized materials, asphalt concrete + graded gravel + ordinary cement concrete, or asphalt concrete + graded gravel + crushed cement concrete.

Cement Concrete Base (Rigid Base) Asphalt Pavement

Composite pavement is a pavement structure with cement concrete [including Jointed Plain Concrete Pavement (JPCP), Jointed Reinforced Concrete Pavement (JRCP), Continuous Reinforced Concrete Pavement (CRCP), steel fiber concrete, prestressed concrete, assembled concrete, and crushed concrete] as the base layer and asphalt concrete as the surface layer. Cement concrete provides high strength and good stability, while asphalt concrete offers comfortable driving and low noise. This composite pavement combines the advantages of both materials, resulting in good performance and durability. Jointed Plain Concrete Pavement (JPCP) and Jointed Reinforced Concrete Pavement(JRCP)-based asphalt pavements may experience reflection cracks at the joints, which can affect performance. Continuous Reinforced Concrete Pavement (CRCP)-based asphalt concrete pavements have good performance and durability because the continuous reinforcement limits the crack width of cement concrete to a certain range (generally less than 1 mm). However, measures must be taken to ensure good adhesion between the asphalt layer and between the asphalt layer and the cement concrete layer.

Cement Concrete Pavement

Cement concrete pavement refers mainly to a pavement structure with cement concrete [including Jointed Plain Concrete Pavement (JPCP), Jointed Reinforced

Concrete Pavement (JRCP), Continuous Reinforced Concrete Pavement (CRCP), steel fiber concrete, prestressed concrete, assembled concrete, and crushed concrete] as the surface layer. Compared with other road materials, cement concrete has high strength, flexural strength and tensile strength. Because of its its high modulus of elasticity, it rigid. Under vehicle loading, the cement concrete layer behaves like a rigid plate, with only small vertical bending and settlement. The pavement structure relies mainly on the bending and tensile strength of the cement concrete slab to bear the vehicle loads, and the pressure transferred to the foundation is much lower than that of flexible pavements due to the slab's stress distribution.

1.3 Road Characteristics and Functions

Over the past 100 years, the rapid development of smart transportation has relied on a series of characteristics of roads and their functions. In contemporary transport systems, however, the functions of roads are no longer limited to providing physical passage and basic traffic capacity. Recent studies indicate that road infrastructure is increasingly expected to support multiple objectives simultaneously, including safer mobility, network resilience under extreme weather and climate change, and more sustainable system operation[19, 20. In this sense, the functional development of roads is shifting from conventional connectivity and traffic service toward integrated roles in safety assurance, risk reduction, climate adaptation, and long-term infrastructure performance management[21]. Compared with other modes of transport, it has the following attributes and features:

1.3.1 Basic Road Attributes

Road construction and road transportation are a form of material production, so the basic attributes of material production, that is, the means of production, labor and labor tools, and the fact that roads exist as material products must be considered. Moreover, roads also have their own unique basic properties.

(1) Public welfare

Roads are widely distributed and cover a wide range of areas, benefiting society and receiving broad attention and support. In particular, because road transportation has played an important role in promoting the development of the social commodity economy in recent years, road construction has received increasing attention.

(2) Commodity

Road construction involves material production. A road is a product and must have the basic attributes of commodities, possessing both commodity value and use value.

This attribute forms the basis for the current development of commercialized roads (also known as toll roads).

(3) Advancement

The advancement of roads mainly refers to their leading role. Roads serve the development of the national economy and society. As a linking connecting industrial and agricultural production and a diver of economic development, their development speed should be higher than that of other sectors. This is commonly referred to as the principle of the "priority development".

(4) Reservation

Road transportation is a capital- and technology-intensive industry, and it is a national capital construction project. Road construction must not only meet current traffic capacity requirements but also consider future increases in traffic capacity; that is, a certain reserve capacity must be ensured. Prior to construction, unified planning, feasibility studies, thorough economic and traffic investigations, enhanced traffic forecasting, and careful design are needed to meet the needs of long-term development.

1.3.2 Economic Characteristics of Roads

As a special material product, roads also have several economic characteristics, including the following:

(1) Roads are linear structures fixed over a wide area and cannot be easily moved

This is different from general industrial production and construction. In industrial production, the production equipment is generally fixed, and the products flow from raw materials to finished products during the production process, whereas road construction is the opposite. The construction industry is similar, but its products are distributed at discrete points rather than a linear manner. Therefore, road construction features a larger operations space, less fixed work locations, greater impacts on the social and natural environment, and higher professional requirements.

(2) Roads have long production cycle and service life

Usually, it takes two or three years to build a road hundreds of kilometers long. High-grade roads are relatively longer, and considerable manpower, material and financial resources are consumed during construction. After being put into use, the service life is generally 10–20 years. During operation, regular maintenance, repair and management are also required.

(3) Although the road is a material product, it does not take the form of a commodity

In the commodity economy, products generally take the form of commodities for exchange in the market and are sold for consumption. After the road is completed, it cannot be sold as a commodity; there is no equivalent exchange through buying and selling, and it is only provided for social service. Its investment costs are recovered through user charges, tolls, or other public financing mechanisms.

(4) Roads have a special consumption process and method

In general, commodity production and consumption are separated in time and space. That is, a commodity must be produced before it can be transported to the market for exchange and consumption. Roads can be constructed and used at the same time, and can be maintained, repaired and reconstructed during use. Production and consumption are inseparable and repetitive in time and space. In the form of consumption, a road is not consumption once but multiple times. This places particularly high demands on road quality to ensure the safe, fast, economical and comfortable vehicles operation under repeated use (consumption).

(5) Complete roads system serves society and the economy

A road is a complete system composed of routes, subgrades, pavements, bridges and culverts, and tunnels. The road network of a region is composed of many roads forming an organic network system. This system functions as a subsystem of the transportation system, requiring the construction of various roads to be planned, and coordinated to serve society and the economy effectively.

1.3.3 Functions of Roads

The main functions of the road include the following three perspectives:

(1) Medium- and short-distance transportation tasks (within 50 km for short-distance transportation and 50 ~ 200 km for medium-distance transportation).
(2) Supplement and connect with other modes of transportation and undertake bulk transportation tasks (such as train and ship transportation).
(3) Under special conditions, independently perform long-distance transportation tasks. With the development of expressways, the role of roads in medium- and long-distance transportation will gradually increase.

The main functions of urban roads include the following six perspectives:

(1) Connect all parts of the city, provide various traffic services, and undertake the transfer and distribution of the city's external traffic.
(2) Form the skeleton of the urban structure and determine the urban pattern.
(3) Provide venues for air defense, fire prevention, earthquake resistance and greening.
(4) Serve as the main channel for laying various public facilities in cities.
(5) Provide ventilation and lighting for the city and improve its living environment.
(6) Divide urban blocks, organize buildings along streets, and express the urban construction style.

1.4 Components of Roads

1.4.1 Main Components of Highways

A highway is a linear structure that includes two components: linear and structural.

Alignment

Highway alignment refers to the spatial geometry and dimensions of the highway centerline. This spatial linear projection is projected on the horizontal, vertical and cross-sectional planes and drawn into graphics reflecting its shape, position and size, namely the plan, longitudinal and cross-sectional views of the highway. In highway design, the horizontal, vertical and cross-sectional aspects influence, restrict, and cooperate with each other, which should be considered comprehensively in the design.

The plane alignment is composed of basic linear elements such as straight lines, circular curves and spiral curves. The profile alignment consists of basic elements such as straight lines (straight slope segments) and vertical curves. The cross- section is composed of different elements, such as lanes, shoulders, medians, curbs, side-walks, and green belts. The requirements of technical economy and aesthetics must be considered in the design of highway alignment.

Structure

A highway structure is influenced by loads and natural factors. It includes roadbeds, pavements, bridges and culverts, tunnels, drainage systems, protection structures, special structures and traffic service facilities. Different grades of roads have different components under different conditions. For example, parking lots are not.

(1) Subgrade

A subgrade is the foundation of the driving part. It bears the driving load transmitted from the road surface. It is an earth structure built of soil and stone according to the route position and certain technical requirements.

(2) Pavement

A pavement is a structure layered on a roadbed with various road construction materials or mixtures to allow vehicles to travel. It directly bears the traffic loads and natural factors, ensuring vehicles can drive safely and comfortably at a certain speed.

(3) Bridge and Culvert

Bridges are structures built to cross natural or artificial obstacles, such as rivers and valleys. Culverts are small drainage structures passing under roadways to vent surface water. On low-grade roads with low water flow, a permeable roadbed with large stones or pebbles, allowing water to pass through wide, shallow rivers that are usually dry or have little flow. Water-passing roads, ferry crossings, wharves, and similar facilities can also be provided at intersections without bridges.

(4)   Drainage System

To prevent natural water, such as surface water and groundwater, from eroding or scouring the subgrade and to ensure its stability, a drainage structure is needed. In addition to the bridges and culverts, side ditches, intercepting ditches, drainage ditches, drops, rapids, blind ditches, seepage wells and aqueducts may be provided. These structures form an integrated drainage system to reduce or eliminate various types of water damage to roads.

(5)   Tunnel

A tunnel is a structure built to allow roadsto pass through the ground or underwater. Tunnels can shorten road distances and allow smooth and rapid travel.

(6)   Protection Structure

Filling slopes, masonry slopes, retaining walls and foot guards built on steep hillsides or roadbed slopes along the river can reinforce roadbed to ensure stable structures. Snow fences, snow shelters, and similar facilities can be installed on road sections prone to snow damage. Additional facilities should be provided to control wind erosion and modify conditions of sand transport and accumulation in the sand-damaged road section. Diversion structures, such as grid dams, spur dams and dams and other indirect protection works, can be installed along the subgrade adjacent to rivers.

(7)   Special Structure

In mountainous terrain and complex geological sections, overhanging road platforms, mid-mountain bridges and stone-proof corridors can be built to ensure the continuity of the road and the stability of the roadbed.

(8)   Roadside Facilities

Some facilities for ensuring traffic safety, management, service and environmental protection along highways include lighting equipment, traffic signs, traffic markings, guardrails, toll stations, signal facilities, monitoring systems, sound barriers, isolation fences, gas stations, public transport stops, parking lots, rest facilities and greening and beautification facilities.

1.4.2   Main Components of Urban Roads

Urban roads connect the main parts of a city, such as residential areas, city centers, industrial areas, stations, wharves and other parts, to form a complete road system, which is usually consists of the following parts:

(1) Motorized lanes and nonmotorized lanes.
(2) Sidewalks and crosswalks.
(3) Intersections, interchanges, pedestrian squares, parking lots, and bus stops.
(4) Traffic safety facilities, pedestrian tunnels, pedestrian bridges, lighting equipment, guardrails, signs, markings, signal lights, etc.
(5) Drainage system, including street ditches, rainwater inlets, manholes and rainwater pipes, etc.
(6) Streetside facilities such as lighting poles, utility poles, mailboxes and fire hydrants, etc.
(7) Various underground pipelines.
(8) Green belts, including central green belts, side green belts, basic green belts, street trees, etc.
(9) Large cities also have underground railways, viaducts, etc.

The main components of road engineering are routes, roadbeds (including drainage systems and protection works, etc.) and pavements. In road design, they are interconnected and influence each other. The route design should not only have an economical and reasonable alignment but also fully consider the natural and topographical factors of the passing area to ensure the stability of the roadbed. The subgrade must have sufficient strength and stability to ensure the overall strength and stability of the pavement structure and ensure safe and rapid driving.

1.5 Scope of Road Engineering

Road engineering aims to provide fast, safe, comfortable and economical road facilities for transportation systems. With respect to this goal, road engineering covers planning, design, construction, maintenance and operation management. Usually, bridges, culverts, and tunnels belong to the categories of bridge engineering and tunnel engineering, respectively, whereas traffic control and management facilities belong to the category of traffic engineering. Therefore, these structures and facilities are beyond the scope of the road engineering course. however, road engineering has evolved from a discipline focused primarily on construction and service performance to one that also addresses digitalization, sustainability, climate resilience, recycling, and life-cycle management. Accordingly, modern road engineering places increasing emphasis on intelligent monitoring, low-carbon pavement technologies, adaptation to extreme weather, and long-term infrastructure performance[21, 22].

1.5.1 Planning

In terms of planning, the purpose of road planning is to determine the problems and deficiencies of the existing road networks and road facilities through assessments of current conditions and transportation demand forecasts. This help to formulate reasonable development or improvement goals, propose appropriate countermeasures and strategies, and implement plans to accommodate future needs. The main contents of road planning include the following:

(1) Investigate the status of the existing road network and road facilities (including the physical condition of the facilities, traffic operation status, economy, etc.), collect the economic and social datas of the service area, and analyze the datas to evaluate the adaptability of the road networks and road facilities to the existing transportation and traffic demands.

(2) Predict the future transportation and traffic demands of the road networks and road facilities according to the forecast and analysis of the economic and social development of the served area, and evaluate the adaptability of existing road networks and road facilities to future transportation and traffic demands.

(3) Formulate development or improvement goals for the road networks and road facilities to meet future transportation and traffic needs, and propose several planning schemes.

(4) Conduct performance analysis of road networks and road facilities for each planning scheme, carry out both individual and comprehensive evaluations, and formulate implementation plans for the selected options.

1.5.2   Design

Route design is usually referred to as geometric design, and its task is to design a safe, comfortable and economical road according to the design speed, traffic volume and service level requirements, as well as the driving characteristics and vehicle operating characteristics. The contents of the route design are as follows:

(1) Based on the functional and technical level requirements of the road, route alignment, control points (must pass locations), bridge sites and tunnel portals are selected following field surveys of physical, economic, topographic, geological, hydrological, and meteorological conditions.

(2) According to the terrain, geological, hydrological conditions along the route, as well as technical standards, the route alignment is selected between the specified control points, determining the geometry of the horizontal, vertical, and cross-sectional elements, as well as intersection design.

The basic requirements for the subgrade are good overall stability and small permanent deformation. The content of the subgrade design mainly includes the following:

(1) Design the cross-sectional shape and slope of the subgrade according to the height and width of the embankment or excavation determined by the route design, considering the rock, soil and hydrological conditions along the route.

(2) According to the local climate, geology and hydrology, the slope stability of the subgrade for high embankments and deep excavation should be analyzed, and retaining structures should be designed when the stability is insufficient.
(3) Analyze the stability and settlement of the subgrade located on the weak foundation, and select appropriate foundation reinforcement treatment measures when necessary.
(4) Where subgrade slope may be peeled, chipped or easily eroded by water, appropriate slope protection measures should be selected.

The basic requirements for the pavement are sufficient bearing capacity, smoothness, skid resistance and low noise. The task of pavement design is to provide a pavement structure that meets service performance requirements during the design life at the lowest life cycle cost. The contents of pavement design mainly include the following:

(1) According to the design period, fuctional requirements, the local natural environment (temperature and humidity), subgrade support conditions, and material availability, propose a selection and combination scheme of pavement structure types and levels.
(2) According to the property requirements of the selected materials as well as the local environmental conditions (temperature and humidity), the mixture composition of each structural layer is designed.
(3) Apply the mechanical model with corresponding calculation theories or empirical methods, determine whether the axial load and the thickness of each structural layer are affected by the environmental conditions and design life.
(4) Comprehensively consider factors such as investment, construction, maintenance and service performance, analyze the life cycle cost of each alternative design, and select the most cost-effective design.

The purpose of drainage design is to quickly remove the surface water falling within the road boundary and drain the surface water and groundwater from the upper side of the road to the lower side, preventing the subgrade and pavement structure from saturation, erosion, or other damage. It involves the design and installation of various surface and underground drainage facilities, such as ditches, pipes, seepage ditches, drainage layers, which are integrated into the road drainage system.

1.5.3 Construction

The purpose of construction is to realize the intention of the design to build road facilities (mainly roadbed and pavement structures) that meet the quality indicators and the performance requirements for intended use. The main contents of the construction are as follows:

(1) Carry out organizational, technical, material and onsite preparations before construction, including team implementation and training, joint review and

onsite verification of design drawings, alignment restoration and construction measurement, and preparation of construction organization plan.

(2) Carry out subgrade earthwork (excavation, transport, filling, compaction and trimming) and conduct foundation reinforcement treatment, drainage structures, support structures, slope protection, etc.

(3) Construct pavement cushion, subbase, base and surface layers, including mixing, transporting, paving, rolling, shaping and curing of mixtures.

(4) Carry out construction management according to regulations and schedule requirements and supervise, inspect and accept construction quality.

1.5.4   Maintenance and Operation Management

During use, road facilities are gradually damaged due to the sustained traffic loads and natural factors. To maintain the functional performance and ensure they meet operational requirements, regular monitoring and condition assessments must be conducted to inform maintenance planning. For road facilities that may be damaged or do not meet operational requirements, maintenance, repairs or reconstruction should be carried out according to the maintenance plannings and maintenance specifications to slow deterioration rates and restore or improve performance[23].

Discussion, Questions and Exercises

(1) The road construction techniques and purposes during the Roman Empire and ancient China were compared. How did these roads reflect the priorities and challenges of their respective civilizations?

(2) Explain the differences between flexible base asphalt pavement, semi-rigid base asphalt pavement, and cement concrete pavement. In what scenarios would each type be most effectively applied?

(3) The unique economic characteristics of roads as material products were discussed. How do these characteristics influence the planning and funding of road projects?

(4) The primary differences in the design and function of urban roads compared with highways were highlighted. How do these differences reflect their distinct roles in transportation systems?

(5) Identify key challenges in designing subgrades for roads built on weak foundations or steep slopes. What measures can engineers take to address these challenges?

(6) Explain why effective drainage systems are critical for road performance and longevity. What are the key components of a road drainage system, and how do they work together?

(7) Analyze how regular maintenance contributes to extending the lifespan and usability of road facilities. The types of maintenance activities typically performed and their importance were discussed.

(8) Reflect on how lessons from historical road developments (e.g., Roman roads or ancient Chinese roads) can inform modern road engineering practices, particularly in terms of sustainability and durability.

References

1. ZHANG J. Road Engineering Monograph [M]. Beijing: Science Press, 2010
2. WU H. Road Engineering Philosophy [M]. Beijing: China Communications Press, 2013
3. HUANG X. Road Base and Pavement Engineering [M]. 6 ed. Beijing: China Communications Press Co., Ltd., 2019
4. Yoder, Eldon Joseph, and Matthew W. Witczak. Principles of pavement design. John Wiley & Sons, 1991
5. Huang, Yang Hsien. Pavement analysis and design. Vol. 2. Upper Saddle River, NJ: Pearson/Prentice Hall, 2004
6. Britannica, Encyclopaedia. "The editors of encyclopaedia." Great Leap Forward (2020)
7. Bryce J, Brodie S, Parry T, Lo Presti D. A systematic assessment of road pavement sustainability through a review of rating tools. Resources, Conservation and Recycling. 2017;120:108–118
8. Santero NJ, Masanet E, Horvath A. Life-cycle assessment of pavements: reviewing research challenges and opportunities. Journal of Cleaner Production. 2016;112:4360–4370
9. Liu S, Shukla A, Nandra T. Technological, environmental and economic aspects of asphalt recycling for road construction. Renewable and Sustainable Energy Reviews. 2017;75:879–893
10. Jacobs JM, Pregnolato M, Qiao Y, et al. Climate change impacts on roadways. Nature Reviews Earth & Environment. 2025;6(9):555–573
11. Holtz, R., C. Barry, and R. Berg. "Federal Highway Administration." (1998)
12. Chen, Anze, et al., eds. Dictionary of geotourism. Singapore: Springer Singapore, 2020
13. National Bureau of Statistics of China. China Statistical Yearbook 2024. Beijing: China Statistics Press; 2024
14. Pompigna, Andrea, and Raffaele Mauro. "Smart roads: A state of the art of highways innovations in the Smart Age." Engineering Science and Technology, an International Journal 25 (2022): 100986
15. Liu, Zhuhuan, Romain Balieu, and Niki Kringos. "Integrating sustainability into pavement maintenance effectiveness evaluation: A systematic review." Transportation Research Part D: Transport and Environment 104 (2022): 103187
16. Wang T, Forgie V, Haynes K. Climate change research on transportation systems: Climate risks, adaptation and planning. Transportation Research Part D: Transport and Environment. 2020;87:102553
17. Thompson, Roger, Rodrigo Peroni, and Alex Visser. Mining haul roads: theory and practice. CRC Press, 2019
18. Fao, F. A. O. S. T. A. T. "Food and agriculture organization of the United Nations." Rome, URL: http://faostat. fao.org 403 (2018)
19. Vecino-Ortiz, Andres I., et al. "Saving lives through road safety risk factor interventions: global and national estimates." The Lancet 400.10347 (2022): 237–250
20. Sias, Jo E., et al. "Climate change impacts on roadways." Nature Reviews Earth & Environment 6.9 (2025): 555–573
21. Liu, Kai, et al. "Global transportation infrastructure exposure to the change of precipitation in a warmer world." Nature Communications 14.1 (2023): 2541
22. Salehi, Safoura, et al. "Sustainable pavement construction: A systematic literature review of environmental and economic analysis of recycled materials." Journal of cleaner production 313 (2021): 127936

23. Hasibuan, Gina Cynthia Raphita, et al. "A Global Perspective on Road Condition Assessment and Maintenance: Trends in Research and Technology Integration." Transportation Research Interdisciplinary Perspectives 36 (2026): 101853

Chapter 2
Road Survey and Alignment Design

A road is a three-dimensional spatial structure with a linear (belt-like) dorm, which mainly includes subgrades, pavement, bridges and culverts, tunnels, etc. Road design is divided into geometric design and structural design. Geometric design is the study of road spatial geometry and falls within the scope of this chapter. Structural design is the study of various engineering components of the road and is covered in Chap. 4. The structural design is based on the geometric design, and the geometric design must consider structural requirements.

Road route selection involves selecting a road centerline that is technically feasible and economically reasonable, and that meets the use requirements based on the basic trend and technical standards of the route, combined with local topography, geology, land features, and construction conditions along the route. Route selection is an important part of road alignment design, and its quality directly affects road performance and project cost. Natural environment, social and economic conditions, alignment technical indices indexes, and other factors should be considered in route selection. Consequently, route selection is a highly technical and policy-driven process influenced by a multitude of complex factors.

2.1 Procedures and Technical Standards

2.1.1 Engineering Feasibility Study

The project feasibility study is an important component of the preliminary work of basic construction, an integral part of the basic construction procedure, and a reliable basis for scientific engineering decision-making. The purpose of the project feasibility study is to comprehensively examine the necessity, technical feasibility, economic rationality, and implementation viability of the project; recommend the optimal scheme; carry out investment estimation and economic evaluation; and

© Tongji University Press Co., Ltd. 2026

H. Li et al., *Road Engineering*, https://doi.org/10.1007/978-981-95-6659-4_2

provide a scientific basis for the decision-making and approval of the construction project, and the preparation of the design assignment. The feasibility study of highway engineering generally includes the following contents [1]:

(1) General/overview. This section includes the basis of construction task and the historical development background, the research scope and main contents of the study, as well as the main conclusions, existing problems, and suggestions.
(2) Evaluation of existing highway technical conditions. These include the current situation and existing problems of the regional transport network, the status and role of the proposed project within the network, and the technical condition and adaptability of existing roads.
(3) Forecast of economic and traffic development. This includes the economic characteristics of the project area, the relationships between economic development and highway traffic volume, and traffic volume forecasting.
(4) Construction scale and standards. These include the project scale, adopted classification (grade), and main technical indicators.
(5) Comparison and selection of construction conditions and schemes. This includes investigation of natural and social conditions along the route, scheme formulation and comparison, proposal of the recommended scheme alignment identification of major control points, and project overview. Environmental impacts should also be analyzed and an environmental impact assessment report prepared.
(6) Investment estimation and fundraising. These include quantities of major works, total investment in highway construction and demolition, source of funds and financing methods. For loan-based or investment- attraction projects, interest rates, repayment methods, and feasibility should also be analyzed.
(7) Project implementation plan. These include plans and requirements for survey and design, construction organization, and the training of project management and technical personnel.
(8) Economic evaluation. These include the determination of economic parameters such as transportation cost, estimation of direct economic benefits and costs, sensitivity analysis, and analysis of indirect economic benefit. The financial evaluation should also be conducted for loan-based projects.

According to the above research results, through comprehensive analysis and evaluation, an optimal construction scheme with advanced technology, lower investment, and better overall benefits is proposed.

2.1.2 Survey and Design Assignment

The survey and design work before highway construction is carried out in accordance with the approved design assignment (or letter of authorization). The design assignment shall be prepared according to the approved project feasibility study report. The design assignment shall be issued by the competent department that proposes the plan or prepared by a subordinate unit and then submitted for approval. The basic contents of the design assignment include the following:

(1) Construction basis, purpose and significance;
(2) Construction scale and nature;
(3) Basic orientation and main control points of the route;
(4) Engineering technical standards and main technical indicators;
(5) Design stage and the completion time of each stage;
(6) Construction period and investment estimation: for phased construction projects, the construction scale and investment estimation of each phase shall be proposed;
(7) Principles of construction organization;
(8) Attached route diagram. In addition, the project quantity, primary construction materials (steel, timber, and cement), and the investment are listed only in the task statement for reference during approval.

After the design assignment is approved, any changes to the main contents, such as construction scale, technical grade, route alignment, and basic direction, shall be subject to approval by the it original approving department.

2.1.3 Design Stage and Its Contents

Design Stage

According to the "Documentation Methods of Highway Engineering Capital Construction Project Design" stipulated by the Ministry of Transport of the People's Republic of China, highway engineering construction projects may adopt one-stage, two-stage or three-stage design methods.

The one-stage design refers to construction drawing design only and is applicable to small-scale projects with simple technology and a clear scheme.

The two-stage design, also known as "preliminary design and construction drawing design", is applicable to general construction projects.

The three-stage design, also known as "preliminary design, technical design and construction drawing design", is applicable to projects with complex technology, insufficient basic data, or special structures such as large bridges, interchanges, and tunnels.

The Main Contents of Each Design Stage

(1) Preliminary design

The preliminary design in the two-stage and three-stage design should be prepared according to the approved feasibility study report, design assignment (or survey and design contract) and preliminary survey data. The purpose of the preliminary design stage is to determine the design scheme. The main contents include reporting the construction principle, selecting the design scheme, calculating the quantities of works and main materials, proposing the construction scheme, preparing the design budget, and providing the text description and chart data. In the preliminary design, when the scheme is selected, the route control points and scheme are checked on site to solicit the opinions of local governments and construction units along the route, and the route layout scheme is essentially implemented. Generally, paper on alignment should be carried out and combined with field surveys to implement and set out necessary control line position piles. Generally, two or more schemes are selected for the route, interchange, tunnel, super large bridge and bridge locations in complex and difficult sections to carry out the survey and set out work and scheme comparisons with the same depth and accuracy, and the recommended scheme is proposed.

(2) Technical design

The technical design in the three-stage design should be prepared according to the approved preliminary design and location survey data. The purpose of the technical design stage is to further implement the design scheme for major and complex technical problems. The main contents include an in-depth exploration, analysis and comparison through scientific experiments and special studies; solving problems to be solved in the preliminary design; implementing technical schemes; calculating project quantities; proposing revised construction schemes; and revising design estimates.

(3) Construction drawing design

The construction drawing design in the two-stage design should be prepared according to the approved preliminary design and location survey data. The construction drawing design in the three-stage design should be prepared according to the approved technical design and supplementary location survey data.

The purpose of the construction drawing design stage is to carry out a detailed design on the approved recommended scheme to meet the construction requirements. The main contents include specifying and deepening the approved construction principles, design schemes and technical decisions; finally determining the quantities of various projects; putting forward written descriptions, charts and materials suitable for construction needs and construction organization plans; and preparing the construction drawing budget.

The one-stage construction design should be prepared according to the approved feasibility study report, design assignment and location survey data. Its purpose and content are to draw up the construction principle, determine the design scheme and project quantity, put forward the text description, chart data and construction organization plan, prepare the construction drawing budget, meet the approval requirements and meet the construction needs.

2.1.4 Technical Standards for Road Survey and Design

The main technical basis for road survey and design includes the following:

- Technical Standard of Highway Engineering (JTG B01-2014)
- Design Specifications for the Highway Route (JTG D20-2017)
- Code for Design of Urban Road Engineering (CJJ37-2016)

  The relevant technical basis for road survey and design includes the following:

- Specifications for Highway Reconnaissance (JTG C10-2007)

  Another technical basis for road survey and design includes the following:

- Documentation methods for designing highway engineering capital construction projects
- Code for Transport Planning on an Urban Road (GB 50220-95)
- Code for Design of Mine Road (GBJ 22-87)
- Design Specifications of Highway Environmental Protection (JTJ/T 006-98)

2.2 Road Design Process

A road is a strip-shaped 3D space entity. In general, a route refers to the spatial position of the centerline of the road, which is a spatial curve composed of straight lines, circular curves and transition curves. This spatial curve is a three-dimensional linear shape determined by its length, height and width; if the time factor is considered, it is a four-dimensional line closely related to the driving speed. The projection of the route on the horizontal plane is called the plane of the route. The vertical section is cut along the road centerline and then expanded, which is called the profile of the route. The normal section of any point on the road centerline is called the cross section of the road at that point. The plan, profile and cross section are the geometric components of the road.

Route design refers to the determination of the spatial position of a route and the geometric dimensions of each component. To facilitate design and research, route design is generally divided into horizontal alignment design, vertical alignment design and cross-sectional design. Horizontal alignment design is the process of studying the basic trend and alignment of roads on the route plane. Vertical alignment design is the process of studying the road longitudinal slope and slope length on the route profile map. Cross-sectional design is the process of studying the subgrade cross-sectional shape on the route cross-sectional map. The three are interrelated and should be designed separately and considered comprehensively. The route positions of roads and urban roads are restricted by natural, social and technical conditions. The purpose of route design is to design an economical and comfortable route with certain technical standards and to meet driving requirements on the basis of investigations, research and the mastery of many materials.

2.2.1 Horizontal Alignment Design

Basic Requirements for Horizontal Alignment Design

(1) Vehicle trajectory

The road is used mainly for vehicles to drive, so it is the basic subject of road design to study the driving law of vehicles. In the horizontal alignment design of a route, the main focus of research is the driving track of vehicles. Only when the horizontal alignment is consistent with or close to the vehicle running track can smooth, comfortable and safe driving be ensured; especially in the case of high-speed driving, research on vehicle running tracks is more important. Many observations and research results show that the track of a moving vehicle has the following geometric characteristics.

- The track is continuous. The running track of the vehicle is continuous and smooth; that is, there is no wrong head, break point or discontinuity at any point.
- The curvature of the track is continuous. Its curvature is continuous; that is, there are no two curvature values at any point on the trajectory.
- The curvature changes continuously. The change rate of its curvature in mileage or time is continuous; that is, the value of two curvature change rates does not appear at any point on the track.

(2) Elements of horizontal alignment

The angle between the guide wheel (or steering wheel) and the longitudinal axis of the vehicle body has the following three relationships:

- The angle is zero;
- The included angle is constant;
- The included angle is a variable.

The vehicle track corresponding to the above three relationships has the following characteristics, as shown in Fig. 2.1:

- The line with zero angle and zero curvature (the radius of curvature is infinite) is a straight line;
- A line with a constant angle and constant curvature (constant radius of curvature) is a circular curve;
- The line shape with the angle as a variable and the curvature as a variable (radius of curvature as a variable) is the transition curve (cycloid).

The horizontal alignment design of modern roads is composed of a reasonable combination of the above three basic geometric alignments, namely, straight lines, circular curves and transition curves, which are called the "three elements of horizontal alignment". The three elements of horizontal alignment are basic components, and the proportion and frequency of each element are not uniformly specified. All the elements are reasonably used and properly configured to meet the requirements

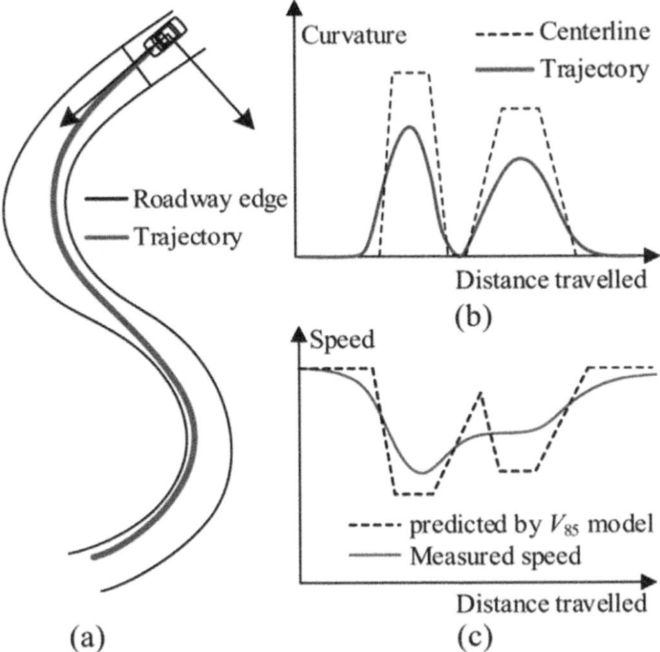

Fig. 2.1 Track and curvature of the vehicle

of vehicle driving. The parameters should be determined according to the terrain conditions, human vision, psychology, road technical level and other conditions (Fig. 2.2).

Straight Line

Straight lines are most widely used in highways and urban roads. When the terrain is flat and there are no large ground obstacles, straight line alignment is often considered first, as shown in Fig. 2.3.

(1) Characteristics of straight lines

The advantages of straight lines are as follows: short and direct route; clear driving direction and good visibility. When a vehicle is running, it has simple force, simple driving operation, and fast driving on the straight line, which can provide better overtaking conditions. The alignment is simple, easy to measure and set, and easy to set other structures.

The disadvantages of straight lines are as follows: the flexibility of linear alignment is poor, and it is difficult to coordinate with the surrounding environment, such as terrain and features; if the length is not used properly, it will not only destroy the

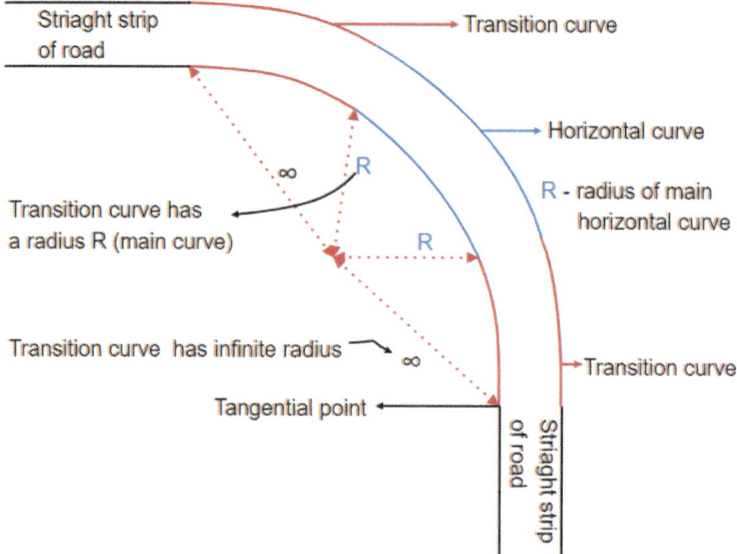

Fig. 2.2 Highway horizontal alignment design example

Fig. 2.3 Real view of a straight line

continuity of the alignment but also inconvenience the coordination of the alignment design itself; a long straight line makes it easy for the driver to feel monotonous, tired and hard to concentrate; and it is difficult to visually measure the distance between vehicles on straight road sections. Traffic accidents are likely to occur due to high-speed driving on long straight roads, and glare is easily caused when driving in opposite directions at night.

(2) Application of straight lines

To better coordinate with the environment, save arable land and reduce the project cost, and under the condition of ensuring the necessary sight distance, the sections with straight line alignment are suitable for:

- A flat area or an open valley between mountains that is not restricted by topography and features;
- Towns and their suburbs, or agricultural areas with square planning and other areas dominated by straight lines;
- Long bridges, tunnels and other structural sections;
- Route intersection and its front and rear;
- A section of a two-lane highway that provides overtaking

Problems to be noted when long straight line alignment is adopted:

- The longitudinal slope on a straight line should not be too large because a long straight line and steep slope are more likely to lead to high speed;
- The combination of a long straight line and a large-radius concave vertical curve is appropriate for easing the stiff straight line;
- When both sides of the road are too open, measures such as planting different trees or setting certain buildings, sculptures and billboards should be taken to improve the monotonous landscape;
- For the horizontal curve at the end of a long straight line or long downhill line, in addition to compliance with regulations such as curve radius, superelevation, and sight distance, safety measures such as setting signs and increasing the antiskid capacity of the road surface must also be taken.

Circular Curves

A curve is a common linear element in plane alignment. Horizontal curves are set for highways and urban roads at all levels regardless of the angle, and circular curves constitute the main part of the horizontal curves. Single curves, complex curves, double intersection or multi-intersection curves, virtual intersection curves and turning curves commonly used in the horizontal alignment of routes generally include circular curves. The circular curve has many advantages, such as easy adaptation to the terrain, good maintainability, a beautiful line shape, and easy measurement and set-out, and is widely used.

It is generally believed that the circular curve has the following main characteristics as the horizontal alignment of the highway:

- The curvature radius R of any point on the curve is constant, and the curvature L/R is constant, so the measurement and calculation are simple;

- Any point on the curve is constantly changing its direction, which is more adaptable to the change in terrain than a straight line is, especially the complex curve composed of multiple circular curves with different radii, which has stronger adaptability to terrain, features and the environment;
- When a car travels on a circular curve, it is subject to centrifugal force and often occupies a greater road width than when it travels on a straight line;
- When the vehicle runs inside a circular curve with a small radius, the sight distance condition is poor, and the sight line is strongly affected by the cutting slope or other obstacles, so traffic accidents easily occur.

The stability of a vehicle running on a curve is affected by the centrifugal force, which is closely related to the curve radius. The smaller the radius is, the more unfavorable it is. Therefore, when the radius of the flat curve is selected, a larger value should be used as much as possible. Only when the terrain or other conditions are limited can a smaller curve radius be used.

When a vehicle is running on a curved road, if the curve is very short, the driver often faces tension when operating the steering wheel, which is dangerous when driving at high speed. In horizontal alignment design, a highway horizontal curve is generally composed of a front and back transition curve and an intermediate circular curve segment curve. To facilitate driving operation, driving safety and skewness, the driving time of the vehicle on any line should not be less than 3 s, and the driving mileage on the curve should be 9 s. If the circular curve in the middle is zero, a convex curve will be formed. However, the convex curve is connected with the two gyratory curves, which is unfavorable for driving. It can only be used in mountain mouths or under special difficult conditions limited by terrain conditions. Therefore, in the design of a horizontal curve, the minimum length of the circular curve should generally meet the travel time of 3 s.

Transition Curves

When the vehicle enters a circular curve from a straight line, the driver should gradually change the steering angle of the front wheel to adapt it to the curve with the corresponding radius. The radius on the straight line is infinite. After entering the circular curve, the radius is R. When the straight line transits to the circular curve, the radius of curvature of the vehicle's travel track is constantly changing. The road section with a changing radius of curvature is the transition curve section. The transition curve is one of the main elements of the road plane alignment. It is a curve with a continuous curvature change between the straight line and the circular curve. On modern expressways, the proportion of transition curves sometimes exceeds that of straight lines and circular curves and becomes the main part of horizontal alignment. Transition curves are also widely used on urban roads.

(1) Effect of the transition curves

- The curvature changes gradually to facilitate the following of the vehicle.

In the process of turning, there is a track line with continuous curvature change. This track line exists objectively regardless of the vehicle speed. Its form and length are related to the vehicle structure, the driving speed of the vehicle, the speed at which the driver turns the steering wheel and other factors. When the vehicle is running at low speed, the driver can use the surplus width of the road surface to keep the vehicle within the lane. It seems that it is unnecessary to set the transition curve. However, when driving at high speed or when the curvature changes sharply, the vehicle may overtake its own lane and drive out of a long transitional track. Therefore, from the perspective of safety, it is necessary to set a route that the driver can easily follow to facilitate the manipulation of the steering wheel.

- Centrifugal acceleration gradually changes to meet passenger comfort requirements.

When the vehicle is running in a straight line, there is no centrifugal force. When a car travels on a circular curve at a certain speed, a centrifugal force is generated, which is inversely proportional to the radius of curvature. When a car drives from a straight line to a circular curve or from a circular curve to a straight line, the sudden change in the centrifugal force due to the change in curvature will produce a lateral impact force on the passengers and make them feel uncomfortable. Therefore, a transition curve should be set to mitigate the change in the centrifugal acceleration.

- The lateral superelevation and widening gradually changed to ensure smooth driving.

There are two significant changes in the transition of a carriageway from a straight line to a circular curve: one is from the double slope section on the straight line to the single slope section on the circular curve, and the second is to change the normal width on the straight line to the widened width on the circular curve. These two changes are completed in the transition curve section to gradually change the transverse superelevation gradient to prevent the vehicle from swinging left and right while driving.

- The best alignment can be obtained when it is properly matched with the circular curve.

A transition curve between a straight line and a circular curve can form a beautiful and visually harmonious linear curve. The route without the transition curve is distorted, as shown in Fig. 2.4a. After the transition curve is set, the route becomes smooth and beautiful, as shown in Fig. 2.4b [2].

(2) Application of transition curves

The transition curve is an important part of the following horizontal alignment combinations:

- Basic type

The basic type is combined in the order of straight line-spiral curve-circular curve-spiral curve-straight line. From the perspective of the overall coordination of the

(a) without a transition curve

(b) with a transition curve

Fig. 2.4 Alignment changes before and after setting the transition curve

alignment, to ensure that the alignment is continuous and coordinated, the length ratio of the spiral curve, circular curves and spiral curve should be designed as 1:1:1 or 1:2:1 when conditions permit, and the length of the circular curves should be 1–2 times that of the transition curves. The value of the spiral curve parameter A meets the requirements of $R/3 \leq A \leq R$.

- S-shaped type

The S-shaped type refers to the combination of two reverse circular curves connected by a spiral curve. The S-shaped type is applicable when the intersection spacing is limited (the intersection spacing is small). The spiral curve parameters A_1 and A_2 of the S-shaped type alignment should be equal and designed to be symmetrical. When different parameters are adopted, the ratio of A_1 to A_2 should be less than 1.5 and not more than 2.0.

- Egg-shaped type

The egg-shaped type refers to the alignment combination of two circular curves in the same direction connected by a spiral curve.

- Convex type

The convex type refers to the alignment type that is connected radially without inserting a circular curve in the middle of two spiral curves in the same direction. The relationship between the angle of the circular curve and the angle of the transition curve is $\alpha = 2\beta_0$ (where α is the angle of the circular curve and β_0 is the angle of the transition curve).

- C-shaped type

The C-shaped type refers to the alignment combination of two spiral curves of the same direction radially connected at the curvature of zero (the curvature of the connection point is 0, $R = \infty$).

- Compound type

The compound type refers to the linear combination of two or more spiral curves of the same direction connected radially at the same curvature.

General Principles of Horizontal Alignment Design

(1) The horizontal alignment should be direct, continuous and suitable, adapt to the terrain and features, and coordinate with the surrounding environment.

In plain and hilly areas with flat terrain, the route can easily be straight and smooth, and the proportion of straight lines is large. However, in mountainous areas or heavy hilly areas where the terrain has great ups and downs, the route is mostly curved. A route with a large proportion of curves must first adapt to the terrain, which is not only an aesthetic problem but also an economic problem and an ecological environment protection problem. The selection and reasonable combination of straight lines, circular curves and spiral curves mainly depend on objective conditions such as terrain and features. It is wrong to unilaterally emphasize that the route should be dominated by straight lines or curves or to artificially specify the proportion of the three.

(2) The balance and consistency of the horizontal alignment are maintained.

To ensure that vehicles on a road run at a uniform speed, each alignment should be continuous without sudden changes in technical indicators. The following points should be noted during design:

- The end of a long straight line cannot be connected with a small radius curve. A long straight line will lead to a high speed. If a small radius curve suddenly appears, it will cause safety accidents due to untimely deceleration, especially at the end of the downhill direction. If it is difficult to avoid a small radius curve due to the terrain, a transition curve with medium curvature should be inserted in the middle, and the longitudinal slope should not be too large.
- There should be a transition between high and low standards. There are transitions between large and small indices on the same grade road, between long straight lines and small radius curves, between adjacent curves with large and small radii, and between sections on the same road designed with different calculated driving speeds.

(3) Alignment with continuous sharp bends should be avoided.

This type of alignment causes inconvenience to the driver and has a negative effect on the comfort of passengers. A sufficient number of straight lines or spiral curves can be inserted between curves during design.

(4) The horizontal curve should have a sufficient length.

If the length of the curve is too short, the driver must turn the steering wheel quickly, which is very dangerous when driving at high speed. Moreover, if a curve of sufficient length is not set to ensure that the change rate of the centrifugal acceleration is less than the fixed value, it is also bad from the psychological and physiological feelings of passengers. When the road angle is small, the curve length is shorter than the actual length, which easily causes the illusion that the curve radius is small. Therefore, it is necessary that the horizontal curve has a certain length.

2.2.2   Vertical Alignment Design

The alignment profile is a section that is vertically cut along the centerline of the road and then expanded. The route longitudinal profile view is a graph that can reflect the shape, position and size of the route on the profile.

There are two main lines on the road longitudinal profile: one is the ground line, which is an irregular broken line drawn according to the ground elevation of each pile point on the central line, reflecting the fluctuation of the ground along the central line of the road; the other is the design line, which is a geometric line with a regular shape determined after a comparison of the technical, economic, aesthetic and other factors. It is drawn by the design elevation points of each stake point on the central line, which reflects the ups and downs of the road, as shown in Fig. 2.5.

The vertical alignment design consists of a straight slope line and a vertical curve design. The straight slope line (uniform slope line or straight slope section) is divided into uphill and downhill parts. Its slope and length affect the driving conditions of vehicles, the economy of transportation and the safety of driving. Some of their critical values and necessary restrictions are determined by the driving performance of passing vehicles. The vertical curve is set for the smooth transition of the line at the slope turning point (slope change point) of the straight line. According to the different forms of slope turning, the vertical curve has two types, concave and convex, and its size is expressed by the curve radius and the curve length (horizontal length), with various configurations of vertical combinations, as shown in Fig. 2.6.

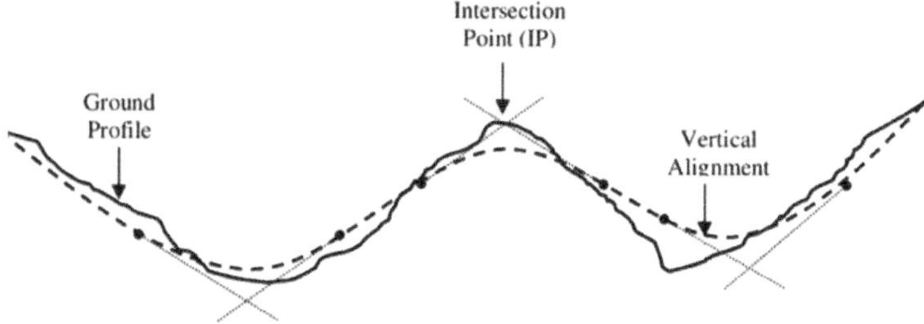

Fig. 2.5   Road longitudinal profile

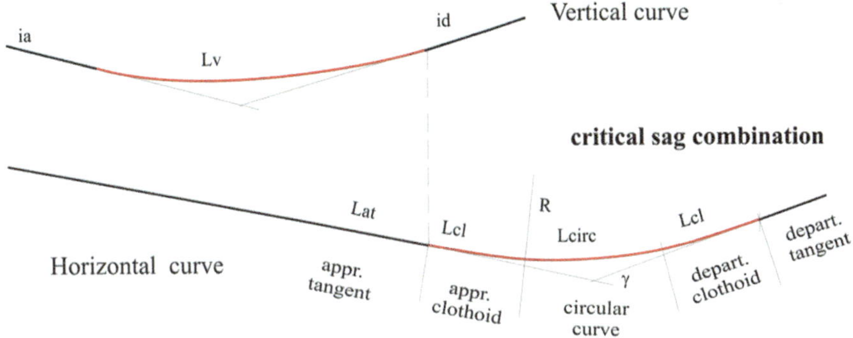

Fig. 2.6 Configurations of vertical combinations

2.2.3 *Combination Design of Horizontal and Vertical Alignments*

The road alignment design starts from the road alignment selection and alignment and is finally reflected in the driver's vision by the three-dimensional alignment composed of horizontal and vertical planes. The combination of horizontal and vertical lines refers to research on how to meet the requirements of visual and psychological comfort and coordination with the surrounding environment on the premise of meeting the requirements of vehicle kinematics and dynamics and ensuring good drainage conditions. Although the horizontal and vertical line shapes are designed according to the above requirements, if the combination of horizontal and vertical alignments is not good, it will not only hinder the exertion of its advantages but also aggravate the disadvantages of both sides and cause danger in driving, and it is impossible to obtain the optimal three-dimensional alignment shape and a reasonable combination of horizontal and vertical alignments, as shown in Fig. 2.7.

Fig. 2.7 Poor combination of horizontal and vertical curves

During the driving process, the actual driving speed is selected by the driver after the three-dimensional alignment is judged. Therefore, the design must not only meet the horizontal and vertical alignment standards but also the visual continuity of the road spatial alignment, with sufficient comfort and safety. When the calculated driving speed is greater than or equal to 60 km/h, it is necessary to pay attention to the reasonable combination of horizontal and vertical alignments and try to achieve continuous alignment, good vision, balanced indicators, a coordinated landscape, safety and comfort. The higher the design speed is, the more comprehensive the factors considered in the alignment design. When the calculated driving speed is less than or equal to 40 km/h, first, on the premise of ensuring driving safety, the alignment element indices (maximum and minimum values) are correctly applied. A reasonable combination of various alignment elements should be achieved as much as possible. Try to avoid and mitigate unfavorable alignment combinations.

The principles of the combination design of horizontal and vertical alignments are shown below [3].

- The driver's sight should be naturally guided visually, and visual continuity should be maintained.

Any alignment that makes the driver feel confused, confused and misjudged must be avoided as much as possible. The ability to visually induce the line of sight naturally is the most basic criterion for evaluating the quality of the combination of horizontal and vertical lines.

- The technical indexes of horizontal and vertical alignment should be balanced. This not only affects the smoothness of the alignment but is also related to the project cost. It is meaningless to adopt a high standard alignment on the plane when the longitudinal alignment fluctuates repeatedly, and vice versa.
- A proper combination of synthetic slopes is selected to facilitate driving safety and pavement drainage.
- Attention should be given to the coordination with the surrounding environment of the road. This approach can reduce driver fatigue and tension and guide the line of sight.

2.3 Route Selection and General Design

The purpose of road alignment is to select the location of the road centerline in the field or on paper according to the nature, task, grade and standard of the road, combined with topography, geology, features and other conditions along the route, taking into account all three factors: flat, vertical and horizontal. The main tasks of road alignment are to determine the road direction and overall layout, the intersection position of specific roads and the elements of selected road curves, and the plane position of the route through paper or field route selection [2].

2.3.1 Principles and Methods

Principles for Road Selection

The route is the skeleton of the road, and the quality of route selection is directly related to the function of the road itself and whether it can play its due role in the road network. It is necessary to comprehensively consider various factors in route selection and properly handle all aspects[4]. Its basic principles are as follows.

(1) In each stage of road design, all kinds of advanced means should be used to conduct in-depth and meticulous research on the route scheme, and the best route scheme should be selected on the basis of multiple scheme demonstrations and comparisons.

(2) On the premise of ensuring safe, comfortable and rapid driving, the route design should be small in engineering quantity, low in cost, low in operating cost, good in benefits and conducive to construction and maintenance.

(3) Route selection should be coordinated with farmland capital construction to occupy less land and try not to occupy high-yield land or cash crop land or pass through economic forest parks (such as rubber forests, tea forests, and orchards).

(4) Roads passing through scenic spots and historical sites should pay attention to protecting the original natural state. Their artificial structures should be coordinated with the surrounding environment and landscape.

(5) During route selection, engineering geology and hydrogeology should be investigated in detail to determine their influences on road engineering[5]. Care should be taken on severe geological sections, and in general, the route should be avoided. When crossing is necessary, the appropriate location should be selected, the crossing range should be narrowed, and necessary engineering measures should be taken. For high-fill and deep excavation subgrade sections, surveying the geotechnical conditions of subgrade slopes and determining the conditions of slopes and basements are essential.

(6) Attention should be given to environmental protection in route selection, especially the impact and pollution caused by road construction and automobile operation.

(7) For expressways and first-class highways, owing to the width of their roads, the characteristics of separation of upper and lower lanes can be utilized according to the terrain, features and natural environment of the areas where they pass.

The above principle of route selection is applicable to all levels of roads. However, when mastering these principles, different levels of roads have different emphases. For example, expressways and first-class highways mainly serve fast and direct traffic between the starting and ending points and important control points in the middle. This function determines that its basic direction should not deviate too far from the general direction. When it needs to be connected with towns along the line, it is advisable to use branch lines. For low-grade places, roads serve mainly local traffic, and they are unable to contact more towns within a reasonable range.

Principles for Road Selection

Many factors influence route selection, some of which contradict each other and others restrict each other. The importance of each factor is different on different occasions, so it is impossible to find an ideal solution at one time. The most effective way is to find the best solution through repeated comparisons and selection in stages, from coarse to fine. Generally, route selection is carried out in three steps according to the work content.

(1) Route scheme selection

Route selection is used to solve the problem of the basic route direction between the starting point and the end point. In this work, we usually first identify various possible schemes from a large area on a topographic map at a small scale (1:100,000 to 1:25,000), collect the relevant information of each possible scheme, make a preliminary selection, and determine several schemes with further comparative value. Then, an on-the-spot investigation is carried out, and the best scheme is obtained through the comparison and selection of multiple schemes. When there is no topographic map, the survey or reconnaissance method can be used to collect data on the spot, and the scheme can be selected. When the terrain is complex or the area is large, the route can be selected by aerial inspection or remote sensing and aerial photography data.

(2) Route belt selection

On the basis of the basic direction of the route, some detailed control points are selected according to natural conditions such as topography, geology and hydrology. These control points are connected to form the route belt, which is also called the route layout. The choice of these detailed control points is still determined by comparison. Generally, the route layout should be carried out on a topographic map at the 1:5000 to 1:1000 scale, and it can only be directly selected on the spot at road sections with simple terrain and a clear scheme.

(3) Specific alignment

After the above two steps, the prototype of the route has been clearly outlined. An alignment is the comprehensive design of horizontal, vertical and horizontal lines in the favorable alignment zone according to the technical standards and route scheme, combined with the relevant conditions and the specific work of determining the road centerline.

2.3.2 Comparison of Route Schemes

Main Factors of Route Scheme Selection

Route design is the most fundamental problem in route design. Whether the scheme is reasonable is not only directly related to the engineering investment and transportation efficiency of the highway itself but also affects whether the route plays its due role in the highway network, that is, whether it meets the national political, economic and national defense requirements and long-term interests.

The following main factors should be comprehensively considered in the route selection scheme:

(1) The significance of the route in politics, economy and national defense; the requirements of national or local construction on the task and nature of the route.
(2) The function of routes in the transportation network system, such as railways, highways and waterways; the relationship with the planning of industrial, mining and towns along the line; and cooperation and land use with the construction of farmland and water conservancy along the line.
(3) The influence of natural conditions such as topography, geology, hydrology, meteorology, earthquakes, etc.
(4) Tourist attractions, historical relics and scenic spots along the route.

The above analysis reveals that many factors influence the choice of route scheme. Many of them are interrelated and mutually restricted. On the premise of meeting the requirements of use tasks and properties, the route should comprehensively consider natural conditions, technical standards and indicators, project investment, the construction period and construction equipment, etc. Through the comparison of multiple schemes and careful selection, a reasonable recommended scheme is proposed.

Route Scheme Selection Procedures

The route is determined through the comparison and elimination of many schemes. The more complicated the natural situation and the greater the distance between the two designated positions, the greater the number of possible comparison schemes and the greater the number of schemes that need to be eliminated. Owing to the limitations of the existing design methods and the natural environment, it is impossible for every route to pass the field survey. Therefore, it is necessary to collect as much existing data as possible. Research and screening are conducted indoors first, and then, the schemes are investigated.

The steps for route selection are as follows:

(1) Collect data. To perform well in highway route selection, we must collect the existing data as much as possible to reduce the workload of investigation.

(2) According to the determined route general direction and highway grade, first, on a topographic map at a small scale (1:1,000,000 or 1:50,000), the various possible routes are preliminarily analyzed. The key points should focus on sections with complex topography, geology and ground features.

(3) Carry out an on-the-spot investigation according to the scheme put forward by preliminary indoor research, together with the new scheme found in the field investigation.

(4) Sort out and summarize the investigation results by item and prepare the project feasibility study report, which provides the basis for the superior to prepare or supplement and modify the design task book.

2.3.3 Route Selection in the Plain Area

Features of the Plain Area

The plain area has a flat terrain and gentle slope. Villages and towns, farmland, rivers, lakes, ponds, swamps, saline soil, etc., are common natural obstacles in plain areas, as shown in Fig. 2.8. Therefore, the main feature of route selection in the plain area is overcoming plane obstacles.

The plain terrain has little restriction on the route. The basic alignment of the route should be short, quick and straight. If there are no residential areas, buildings, factories, mines, nature reserves, scenic spots, etc., between the two control points, a straight line can be adopted. However, in towns and industrial areas with dense farmlands, irrigation channels and dense populations, the route needs to be turned.

Fig. 2.8 Factors controlling the selection of road alignment

Therefore, for route selection in plain areas, the places specified in the general direction of the route, such as cities, factories, farms, villages and towns, as well as cultural relics and scenic spots, should be taken as large control points. Then, field surveys among the large control points are carried out to understand the advantages and disadvantages of farmland and the distribution of ground features and determine which ones can be worn, which ones should be bypassed and how to avoid them to establish a series of intermediate control points. Generally, the route should go from one control point to another without any distortion. To improve the road capacity, it is necessary to coordinate the horizontal and vertical surfaces of the route, and it is also necessary to set appropriate vertical curves at the slope turning point.

In plain areas, the combination of short-term and long-term route schemes should be fully considered. Higher standards should be adopted as much as possible in horizontal and vertical alignment so that the original subgrade, bridges and culverts can be fully utilized when the highway grade is upgraded in the future.

Key Points for Route Selection in the Plain Area

The layout of the route in the plain area should consider mainly politics and economy on the premise of basically conforming to the route direction. The following points should receive more attention.

(1) Correctly handling the relationship between roads and agriculture

Farmland in the plain area is covered with crisscrossing channels. Route selection should focus on supporting agriculture.

Some farmland will inevitably be occupied by newly built roads in the plain area. However, we should try our best to occupy fewer fields. The route should be compatible with the construction of agricultural water resources and intersect as few irrigation channels as possible. When the route direction is essentially the same as the channel direction, it can be laid along the riverbank.

(2) The connection between the route and the town should be considered.

There are more towns and villages, industries and other facilities in the plains, and routes should be selected to avoid them. As far as possible, no or little damage should be done, and higher technical specifications should be used for passage.

High-grade highways should avoid crossing towns, industrial and mining areas and densely populated residential areas, but they should also consider facilitating transportation for agriculture, the masses and links with industrial and mining areas. The local government agreed that the general communication of counties, townships, and villages directly for agricultural transport services on the road also crossed the town and countryside. However, there should be sufficient width and sight distance for pedestrians and traffic to ensure their safety.

(3) Handle the relationship between overpass and bridge location.

Large and medium-sized bridges are often the control points of a route. In the general direction of a route, bridge roads should be considered comprehensively, favorable bridge locations should be selected, and routes should be laid. It is necessary to prevent the difficulty of the bridge span from being increased only by considering the straightness of the route, regardless of the conditions of the bridge location, and to prevent one-sided emphasis on the bridge location so that the route winding is too long and the standard is too low. In general, the midline of the bridge should be orthogonal to the mainstream flow of the flood as much as possible, and the bridge and approach should be in a straight line.

In principle, the locations of bridges and culverts should follow routes. However, in the case of excessive skew (generally when the angle between the bridge axis and the flood flow direction is smaller than 45 degrees) or a river ditch that is too curved, measures can be taken to change the river or route, and the angle between the bridge axis and the flow direction can be adjusted to avoid excessively increasing construction difficulties and project investment.

(4) Paying attention to soil hydrological conditions.

The soil hydrological conditions in the plain area are poor, especially in the river network lake area, where the terrain is low and the groundwater level is high, which makes the stability of the subgrade problematic. Therefore, the route should be laid along the higher terrain close to the watershed as much as possible.

(5) Correctly handle the relationship between the old road and the new road.

When the designed traffic volume is large and new roads need to be built, the relationship between the new roads and the old roads should be handled according to the situation.

(6) As close as possible to the origin of the building materials.

Generally, there is a lack of sand and gravel building materials in plain areas, so the route should be as close as possible to the origin of materials to reduce the cost of construction, maintenance and material transportation.

2.3.4 Route Selection in Mountainous Areas

Features of the Mountainous Area

Mountain areas are characterized by steep slopes and rapid flows, and the terrain is very complex. There are many rocks, thin soils and complicated hydrogeological conditions. However, the mountains in the mountainous area are clear, which highlights the direction of route selection in the mountainous area, either along the mountains or across the mountains.

Key Points for the Valley Line

The valley line is a route arranged along the riverbank, as shown in Fig. 2.9. After the optimization of natural water flow, the riverway could form a gentle slope. Along the banks of streams and rivers, relatively open terraces and beaches are formed, which is a landform suitable for roadways.

Generally, the valley bottoms of rivers in mountainous areas are not wide, the terraces on both sides of the river are narrow, and the valley slopes are slow and steep, sometimes shoals and cliffs. Most rivers are curved, with steep concave banks and slow convex banks. Both sides of the bank are steep cliffs, that is, canyons, and open places often have wide terraces, which are the only good cultivated land in mountainous areas. The geological conditions of river valleys are complex, and diseases such as landslides, rock piles and debris flows often occur. Owing to the lack of sunshine, there is often snow, avalanches and salivate ice in canyons in cold areas. Rivers in mountainous areas usually have low flow, but in the case of heavy rain and flash floods, the torrents often carry sediment, gravel, trees, etc., which rapidly discharge and scour river banks and destroy bridges and culvert, causing considerable losses.

The above natural conditions cause difficulties in route selection, but compared with other routes in mountainous areas, the horizontal and vertical alignments of

Fig. 2.9 Valley line

valley lines are the best. When making full use of favorable environmental conditions, valley lines can outperform other alignments in terms of route standards and engineering costs. This is why river alignments are often the preferred option for mountainous areas.

When selecting a route, we should grasp the main contradiction, combine the nature and grade standard of the route, and solve it according to local conditions.

(1) River bank selection

Owing to the advantages and disadvantages of both sides of the river valley, the topography, geology, hydrology and other conditions on both sides of the river, as well as the factors of irrigation and water conservancy planning, should be compared in route selection. When a bridge is not complicated, to avoid unfavorable terrain and unfavorable geological zones and to reduce mileage, crossing the river and changing banks to set up lines are possible. The choice of riverbank should generally be determined via a combination of many factors, such as technical and economic comparisons.

a. Topographic, geological and hydrological conditions.
b. Bank selection in snowy and frozen areas.
c. The distribution of residential areas, urban and rural construction, industrial and agricultural development and other transportation and water conservancy facilities should be considered.

(2) Route height

According to the relationship between the height of the valley line and the design flood level, there are two types of lines: low lines and high lines.

Generally, a low line refers to a line that is not much higher than the design water level (including the wave height and safe height), and the side slope near the water of the subgrade is often threatened by floods. The advantage of the low line is that the horizontal and vertical line shapes are straight and gentle, and it is easy to achieve higher standards. The quantity of earthwork is small, the slope is low, and it is easy to stabilize. The route has a large range, which makes it convenient to use favorable terrain and avoid bad terrain and geological conditions. It is convenient for crossing tributaries, and it is easy to handle when it must cross the mainstream. The greatest disadvantage is that it is threatened by floods, and many protection projects exist.

The high line refers to the route that is higher than the design water level and is essentially not threatened by floods. Generally, it is used to avoid snow burying and difficult projects. The advantages are that it will not be invaded by floods, and the abandoned earth and stone can be easily treated. Otherwise, the disadvantage is that because the high line is generally located on the hillside, the route will inevitably bend with the mountain, resulting in poor alignment and large projects.

(3) Bridge location selection

According to the relationship between the route and the river, there are two types of bridges: those crossing tributaries and those crossing the mainstream.

The choice of bridge location across tributaries is generally a local scheme problem. However, the choice of bridge location across the main river is mostly a route layout problem. The bridge location across the main river is often the control point used to determine the route direction. It depends on and influences the choice of river banks. In addition to riverbank selection, we should carefully analyze the choice of bridge location across the river. When the route needs to change the bank for route due to topography and geology, if the bridge location is not well selected, it will be forced to cross the river, which will either cause poor alignment of the bridge head or increase the bridge project. Therefore, when the riverbank is chosen, it is necessary to research and address the layout of the bridge site and the route of the bridge head.

Key Points for the Ridge-Crossing Line

The route that climbs up the ridge along the hillside on one side of the watershed, passes through the site entrance at an appropriate place, and then descends along the hillside on the other side is called the ridge-crossing line, as shown in Fig. 2.10. The route needs to overcome great height differences, the terrain and geological conditions are complex, the project is arduous and concentrated, and the length and plane position of the route depend mainly on the arrangement of the longitudinal slope of the route. Therefore, in the route selection of the ridge-crossing line, the route profile should be the leading factor. The main problems to be solved in the layout of the crossing line are the selection of the pass, the elevation of the crossing line and the drawing up of the route on both sides of the pass. They are interrelated and influence each other. The layout should be considered comprehensively, and the relationships among them should be handled well.

(1) Pass selection

The pass is an important control point that reflects the plan of the ridge-crossing line and should be selected in a large range that basically conforms to the route direction. The location, elevation, topography, geological conditions, and line spreading conditions of the pass should be fully considered.

On the premise that the location of a pass is essentially in line with the route, a pass with a low elevation, good conditions for line spreading and favorable topography and geology should be selected. However, in actual terrain, these conditions may not be met at the same time. Therefore, it is necessary to select several passes in different positions at the same time and then determine an optimal scheme through careful comparison. The elevation of the pass and the height difference between the pass and the control point at the foot of the mountain have a direct effect on the length of the route, the amount of engineering work and the operating conditions. Typically, a pass with a lower elevation should be selected. In alpine areas, especially in snow and ice areas, high-altitude routes are bad for driving. Therefore, sometimes, to lower the number of passes, even if the direction deviates somewhat and the distance is somewhat far away, the scheme can be considered. However, if snow and ice are not

Fig. 2.10 Ridge-crossing line

severe, passes with good routes and geological conditions should not be abandoned easily, even if the elevation is slightly high.

The mountain slope spreading line is the main design content of mountain crossing lines. However, the degree of twisting and turning of slopes, the steepness and gentleness of cross slopes, geological conditions, etc., are all directly related to the alignment standards and the amount of engineering work. Therefore, the selection of the pass must be considered in combination with the conditions of the line extension of the slopes on both sides. If there is a hillside with good geological conditions, a gentle terrain, and favorable terrain for spreading the line and lowering the slope, even if the pass position is slightly off or higher, it should be included in the selection range. The general geological structure of a pass is weak, and bad geology often exists. In-depth investigations and studies should be conducted on its stratigraphic structure to determine its nature and influence on highways. The corresponding measures in the design should be taken, with a focus on protection to ensure the smooth passage of the route.

(2)   Selection of mountain crossing elevation

When a route passes through a watershed, that is, over mountains, cuttings or tunnels are often excavated. The lower the elevation of the mountain crossing is, the shorter the route, but the deeper and longer the cutting or tunnel is, the greater the engineering quantity. Therefore, mountain crossing elevation should be determined through technical and economic comparisons in combination with the route grade, topography and geology of the mountain crossing section, the plan of the line spreading on both sides, the method of mountain crossing and other factors. These factors influence each other, so we must comprehensively analyze and study various possible comparison schemes and make a reasonable choice.

The main methods and elevations of mountain crossings are as follows.

a. Shallow excavation and low filling
b. Dig deep into the pass
c. Tunnel crossing

(3) Exhibition line of the route on both sides of the pass

Line spreading is a route method used to overcome the height difference by extending the route length of the terrain to ensure that the longitudinal slope of the route in mountainous areas meets the technical standards.

A. Layout of the line

The elevation of the crossing line is overcome mainly by the spreading line on the slopes on both sides of the pass. Although the hillside terrain is very different and the alignment varies, the layout of the route should first be guided by the longitudinal slope; that is, the combination of the horizontal, vertical and horizontal surfaces should be dominated by the longitudinal profile. The ridge-crossing line is realized by reasonably adjusting the slope and setting the necessary geometric curves by taking advantage of favorable terrain and geology, and the layout of various curves should also be selected according to the longitudinal slope. Only the route scheme that meets the standard of a longitudinal slope can be established. Therefore, the layout of the spreading line must start from the arrangement of the longitudinal slope, and its working steps are as follows:

- Draw up a rough route.
- Try slope route.
- The control points are analyzed and implemented, and the layout plan is determined.

B. Line spreading mode

There are three main ways to spread the ridge line: natural spreading, the switch-back curve and spiral spreading.

The natural spreading line overcomes the height difference along the natural terrain with an appropriate slope. The advantage of the natural stretch line is that it is in line with the basic direction of the route, the journey and elevation rise and fall are unified, and the route is the shortest. Compared with the back stretch line, the line shape is simple, and the technical index is generally greater. In particular, the routes do not overlap, which is beneficial for driving, construction and maintenance. If the area where the route passes is geologically stable and there is no obstacle to split terrain, this scheme should be adopted as far as possible for the route. Its disadvantage is that the degree of freedom to avoid difficult projects or bad geology is not great, and the only way is to adjust the slope. If you encounter high cliffs, deep valleys or large-scale geological diseases, it is difficult to avoid them, so you have to take other ways to spread the line.

When the height difference between control points is large or the natural route cannot meet the demand due to the restriction of terrain and geological conditions, the route can be laid by setting a turning curve in favorable terrain. The disadvantage of the retracing line is that it easily overlaps before and after the line on the same slope. In particular, the upper and lower lines near the back curve are very close to each other, which is unfavorable for driving, construction and maintenance. Therefore, this line spreading method can be adopted only as a last resort. The advantage of the switch-back curve is that it is easy to make use of favorable terrain and avoids bad terrain, geology and difficult projects. The location of turning back is closely related to the size and quality of the turning back curve project, so it should be carefully selected.

When the route is restricted and it is necessary to raise or lower the height in a certain place to make full use of the terrain, the spiral spreading line can be considered. Generally, the spiral spread line is circled by mountain packs on the ridge and crossed by a dry bridge or tunnel. There are also detours in the valley, which use bridges to cross ditches and lines. At present, as a method of highway route selection, spiral route laying can also be regarded as a type of change in turning back route laying. Under certain terrain conditions, it can be used to replace a set of turning curves. Compared with the turning curve, it has the advantages of better alignment and avoiding overlapping routes. However, owing to the need to build tunnels or Gao Qiao and long bridges, the cost is very high, so it is seldom used. When necessary, it should be compared in detail with the plan of turning back the line according to the nature and tasks of the route.

2.3.5 Route Selection in Hilly Areas

Features of the Hilly Area

The hilly area is the terrain between the plain area and mountain area. Its topography is characterized by continuous hills, one after another, tortuous hills, low ridges, wide ridges, gentle slopes and low relative heights. Hilly areas include microhills and heavy hills. The microhilly area has a small fluctuation, the natural slope of the ground is lower than $20°$, the hills and valleys are sparsely distributed, and the relative height difference is within 100 m. The hilly area fluctuates frequently, and the relative height difference is large. The natural slope of the ground is above $20°$. The hills and valleys are densely distributed, and there are deep valleys and high watersheds. The horizontal and vertical parts of the route are limited by the terrain.

With the ups and downs of terrain in hilly areas, the changes in ground features are also great. Generally, agriculture in hilly areas is relatively developed. There are many kinds of land planted areas, such as paddy fields in lowlands, dry land or economic forests in sloping fields, and many small water conservancy facilities. Residential areas, buildings, landscapes, cultural relics and other facilities often appear in flat areas. These locations are the control points that should be considered when selecting routes.

The characteristics of route selection in hilly areas are as follows:

A. There are many local schemes. Because there are many hills and valleys in the hills, the route is flexible. There are generally many feasible route schemes.
B. Coordination and close cooperation among the horizontal, vertical and horizontal lines are needed. Because of the twists, turns and frequent ups and downs of the hilly terrain, the horizontal, vertical and horizontal aspects impose great constraints on each other.
C. The form of the subgrade is mainly half-filled and half-excavated. Because of the terrain characteristics in hilly areas, there is often a certain cross slope on the ground where the route passes, but the cross slope is generally not too steep. There is a large contradiction between the route, agricultural and forestry land and water conservancy facilities.

Key Points for Route Selection in Hilly Areas

The topography of hilly areas is complex, so different route methods should be adopted according to the specific topography of the area where the route passes.

A. Flat areas in a straight line. The terrain between the two control points is flat, and the route is generally oriented in the direction of the plain area. If there are no obstacles such as ground features, geology, scenery, cultural relics, etc., routes should generally be in a straight line. If there are obstacles, etc., an intermediate control point should be added, with a long and slow curve with a small turning point and a large radius as the main one.
B. Slope area—even slope line. The uniform slope line refers to the line connecting the ground points with a uniform slope along the natural terrain between two points. Even a slope line is obtained after several slope tests. When there is no barrier between two control points and other factors, it can be laid directly according to the even slope line; if there are obstacles, an intermediate control point should be set up at the obstacles, and the sections should be controlled according to the uniform slope line.
C. Ups and downs—take the middle. The horizontal slope of the undulating ground is relatively slow, so "walking in the middle" means choosing a reasonable route with a smooth plane and a balanced vertical plane between the even slope line and the straight line.

In other words, many feasible schemes exist, and the ground factors are complicated in route selection in hilly areas. The differences among the schemes are sometimes not obvious, which requires route selection personnel to strengthen the reconnaissance investigation, perform detailed analysis and comparisons by means of sectional routes and gradual approaches, and finally select a reasonable route.

2.4 Applications and Prospects of BIM

The building information model (BIM) was introduced in pilot projects in the early 2000s to support the building design of architects and engineers. Consequently, major research trends have focused on improvements in preplanning and design, clash detection, visualization, quantification, costing and data management[6].

2.4.1 Overview

The Building Information Model (BIM) is defined by international standards as "shared digital representation of physical and functional characteristics of any built object which forms a reliable basis for decisions." BIM contains the precise geometry and relevant data needed to support the design, procurement, fabrication and construction activities required to realize a building. After completion, this model can be used for operations and maintenance purposes.

The use of BIM focuses on preplanning, design, construction and integrated project delivery of buildings and infrastructure. Recently, the research focus has shifted from earlier life cycle (LC) stages to maintenance, refurbishment, deconstruction and end-of-life considerations, especially for complex structures.

BIM can be viewed from a narrow and broader perspective. BIM in a narrow sense comprises solely the digital building model itself in the sense of a central information management hub or repository and its model creation issues (technical issues). Commercial BIM platforms offer integrated data management, component libraries and general functionalities. The widespread differentiations of BIM include 3D (spatial model with quantity takeoff), 4D (plus construction scheduling) and 5D (plus cost calculation) BIM. BIM involves not only using three-dimensional intelligent models but also making significant changes in the workflow and project delivery processes.

2.4.2 Benefits of BIM

The key benefit of a building information model is its accurate geometrical representation of the parts of a building in an integrated data environment. Other related benefits:

a. Faster and more effective processes: Information is more easily shared and can be value added and reused.
b. Better design: Building proposals can be rigorously analyzed, simulations can be performed quickly, and performance can be benchmarked, enabling improved and innovative solutions.

c. Controlled whole-life costs and environmental data: Environmental performance is more predictable, and lifecycle costs are better understood.
d. Better production quality: Documentation output is flexible and exploits automation.
e. Automated assembly: Digital product data can be exploited in downstream processes and used for manufacturing and assembly of structural systems.
f. Better customer service: Proposals are better understood through accurate visualization.
g. Life cycle data: Requirements, design, construction, and operational information can be used in facility management.

2.4.3 Application of BIM to Road Design

There are significant differences between road projects and general construction projects. General construction projects are limited by the terrain, but road projects should be built according to the terrain of the surrounding area, which should achieve not only a high safety coefficient but also environmental protection and promote sustainable development[7].

The processes of applying BIM to road work are as follows.

(1) Establishment of the road information model (RIM)

The basic data of road geometric design and terrain data are used to construct accurate three-dimensional models of the ground, roads, bridges and tunnels, and then satellite or aerial digital images are used to map the model to create a road virtual reality space, as shown in Fig. 2.11.

Fig. 2.11 Road information model (RIM)

(2) Establishment of the 3D road model

The 3D model of a road includes a 3D model of the road, a 3D model of the structure (such as bridges, tunnels, culverts, etc.), and a 3D model of auxiliary facilities (such as road guardrails, mileage road signs, street lights, etc.). The processes of establishment are shown in Fig. 2.12.

(3) Analysis of spatial driving sight distance

Virtual construction technology is used to construct a spatial sight distance measurement model to inspect and verify the sight distance index of newly built/renovated roads[8], as shown in Fig. 2.13.

(4) Geological 3D reconstruction

On the basis of the analysis of the surrounding terrain, the best design scheme (bypass or tunnel) is chosen.

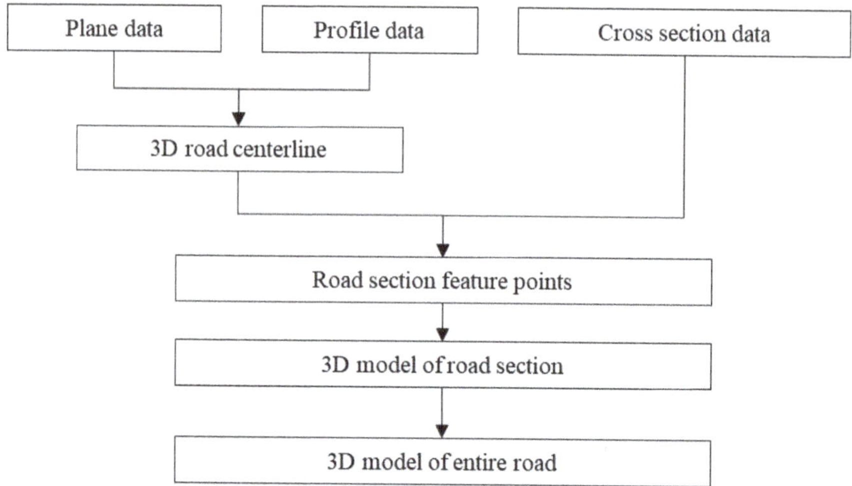

Fig. 2.12 Establishment of the 3D road model

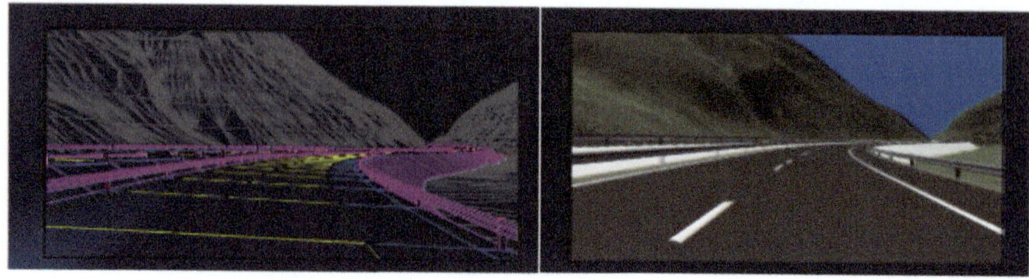

Fig. 2.13 Analysis of spatial driving sight

2.4.4 Prospects of BIM on Roads

Currently, there are certain risks and challenges associated with the application of BIM.

(1) Lack of determination of the ownership of BIM data.
(2) Who controls the data input model and is responsible for any inaccuracies?
(3) Responsibility for appropriate technical interfaces between the various processes?
(4) There is no clear consensus on how BIM should be implemented or used.

Long-term trends, such as increased digitalization and automation, increasing existing building stocks and sustainability requirements, and emerging technologies, such as cloud computing, semantic web technology and mobile BIM devices, will stimulate and extend BIM implementation in existing buildings and infrastructures. However, owing to the revealed state-of-the-art BIM implementation in existing buildings, this area has many challenging future research opportunities.

Discussion, Questions and Exercises

(1) Explain the key components of a highway project feasibility study. Why is it essential to conduct such a study before initiating road construction?
(2) The one-stage, two-stage, and three-stage design methods for highway engineering projects are compared and contrasted. Under what circumstances would each method be most appropriate?
(3) The three main elements of horizontal alignment design (straight line, circular curve, and transition curve) are discussed. How do these elements ensure vehicle safety and comfort?
(4) The challenges and considerations for road route selection in plain, hilly, and mountainous areas should be compared. Examples of strategies used to address these challenges are provided.
(5) The role of geological and hydrological factors in selecting a route for highway construction is discussed. How can poor geological conditions impact the long-term sustainability of roads?
(6) Describe the benefits of applying Building Information Modeling (BIM) in road design and maintenance. What are the potential challenges in implementing BIM for large infrastructure projects?
(7) What are the advantages of using transition curves in horizontal alignment design? How do they contribute to the safety and comfort of high-speed travel?
(8) Analyze how route design can balance economic feasibility and environmental sustainability. Examples of measures taken to minimize ecological damage during road construction are provided.

References

1. YANG S. Road Survey and Design (the third edition) [M]. China Communications Press, 2009
2. LIAO M, SHI G, LI D. Road Survey and Design (the second edition) [M]. Wuhan University Press, 2020
3. XU J. Road Survey and Design (the fifth edition) [M]. China Communications Press, 2018
4. G.S., Kalamaras L., Brino G., Carrieri C., Pline P., Grasso (2000) Application of Multicriteria Analysis to select the best highway alignment Tunnelling and Underground Space Technology 15(4) 415–420. 10.1016/S0886-7798(01)00010-4
5. Dalia, Said Ahmed, Foda Ahmed, Abdelhalim Mustafa, Elkhedr (2024) Performance—Based Route Selection for Mountainous Highways: A Numerical Approach to Addressing Safety Hydrological and Geological Aspects Applied Sciences 14(13) 5844. 10.3390/app14135844
6. Salman, Azhar (2011) Building Information Modeling (BIM): Trends Benefits Risks and Challenges for the AEC Industry Leadership and Management in Engineering 11(3) 241–252. 10.1061/(ASCE)LM.1943-5630.0000127
7. Valeria, Vignali Ennia Mariapaola, Acerra Claudio, Lantieri Federica, Di Vincenzo Giorgio, Piacentini Stefano, Pancaldi (2021) Building information Modelling (BIM) application for an existing road infrastructure Automation in Construction 128103752-10.1016/j.autcon.2021.103752
8. Jaehoon, Jung Michael J., Olsen David S., Hurwitz Alireza G., Kashani Kamilah, Buker (2018) 3D virtual intersection sight distance analysis using lidar data Transportation Research Part C: Emerging Technologies 86563–579 10.1016/j.trc.2017.12.004

Chapter 3
Pavement Materials

Road materials are the foundation for the construction and maintenance of transportation infrastructure, such as roads and bridges, and their performance and types directly influence the road's performance, service life, and structural integrity in road engineering. From low-grade gravel pavement to high-grade asphalt concrete pavement and cement concrete pavement, the progress and development of road materials has supported improvements in highway pavement performance and innovations in pavement structure. With the rapid growth of road traffic and increasing traffic volume and vehicle loads, higher requirements are being placed on the performance of road materials. The scientific and reasonable selection, design, and application of road materials constitute the basis for ensuring the quality of road engineering and improving the technical level of road engineering construction and maintenance.

3.1 Classification of Pavement Materials

3.1.1 Requirements for Pavement Materials

During the service period of road engineering, the effects of traffic loads and environmental factors on the road pavement structure gradually weaken with increasing depth, and the requirements for the strength, bearing capacity, and stability of road materials also decrease progressively with increasing depth. Therefore, the pavement structure is constructed from bottom to top as a multilayer system, comprising a subbase course, a base course, and a surface course. The road pavement structure layers are shown in Fig. 3.1.

The surface course is directly subjected to traffic loading and is significantly affected by temperature and humidity changes in the natural environment. Therefore, the materials used for the surface course structure should possess sufficient strength,

© Tongji University Press Co., Ltd. 2026

H. Li et al., *Road Engineering*, https://doi.org/10.1007/978-981-95-6659-4_3

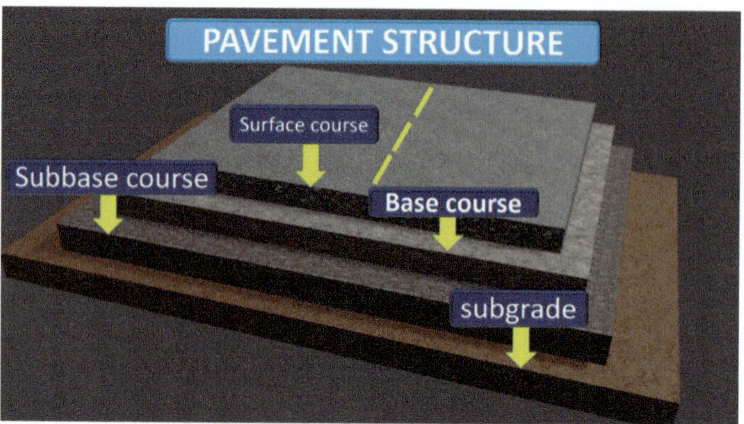

Fig. 3.1 Road pavement structure layers

stability, durability, and good surface characteristics. Commonly used materials include asphalt mixtures, cement concrete, granular materials, and block materials.

The base course is located under the surface course. It primarily bears the vertical stress of the vehicle load transmitted from the surface course and distributes this stress downward into the cushion and subgrade. Therefore, the base course material should have sufficient strength, stiffness, and the ability to disperse stress. Although the effect of environmental factors on the base course is less than that on the surface course, the base course material should still have sufficient water stability and erosion resistance to ensure the stability of the surface course structure. Commonly used materials include binder-stabilized mixtures; crushed stone or gravel mixtures; and natural gravel, concrete and asphalt-stabilized aggregates.

The subbase course is a structural between the base course and the subgrade, and its main function is to improve the humidity and temperature of the subgrade, diffuse the load stress transmitted by the base course, and reduce the deformation of the subgrade. It is usually set up in seasonally frozen areas or sections where the water temperature of the soil base is poor to ensure the strength, stability, and frost resistance of the surface and base courses. Although the strength requirement of subbase material is not high, it should have sufficient water stability, frost resistance, and temperature insulation. Commonly used materials include crushed stone or gravel mixtures and binder-stabilized mixtures.

3.1.2 Classification of Pavement Materials

Commonly used road materials can be classified into the following categories:

(1)   Sand and gravel materials

Sand and gravel materials include artificially crushed rock or rolled crushed stone, natural gravel, and various industrial metallurgical slags with stable performance (such as cinder, blast furnace slag, and steel slag), which are the most widely used materials in road engineering. Among them, larger block stones after processing can be directly used for masonry roads, bridge engineering, and ancillary structures. Rock aggregates with stable performance can be used to make asphalt mixtures or cement concrete, which can be used for paving asphalt pavement or cement pavement, and can also be directly used for paving road bases, cushions, or low-grade road surfaces. Some active mineral materials or industrial waste residues, such as granulated blast furnace slag and fly ash, can be used as raw cement materials or as admixtures in cement, concrete, and asphalt mixtures.

(2) Binders and polymers

Asphalt, cement, and lime are commonly used binders in road engineering, and their role is to bind loose aggregate particles into a whole material with a certain strength and stability. Polymer materials such as plastic (synthetic resin), rubber, and fiber can also be used as binders. In addition to being used as joint fillers for concrete pavement, they can also be used to improve the technical performance of road materials, such as by preparing modified asphalt and producing polymer cement concrete.

(3) Asphalt mixture

An asphalt mixture is a composite material composed of mineral aggregates and asphalt binder, which has high strength, flexibility, and durability. Asphalt pavement has the advantages of comfortable driving, simple maintenance, smoothness, and less dust. It is the most important pavement structure of high-grade highways and urban roads.

(4) Cement concrete and mortar

Cement concrete is a composite material composed of cement and mineral aggregates that has high strength and stiffness and can bear heavy vehicle loads. It is used mainly for bridge structures and high-grade road surface structures. Cement mortar is mainly composed of cement and fine aggregates, which are used in masonry and plastering structures.

(5) Inorganic binder-stabilized mixture

An inorganic binder-stabilized mixture is a composite material formed by lime (fly ash), a small amount of cement (lime), or a soil curing agent is used as a stabilizing material, and the loose soil and gravel aggregates are stabilized and cured. It possesses a certain strength, a plate-like body, and the ability to diffuse stress, but its wear resistance and durability are slightly inferior. Therefore, it is usually used for base courses of high-grade pavements or surface courses of low-grade roads.

3.2 Mineral Aggregate

Aggregates, including artificially crushed rock or rolled gravel, natural sand and gravel, and various industrial metallurgical slags with stable properties, such as cinders, blast furnace slag, and steel slag, are the most commonly used materials in road and bridge engineering structures. After processing, larger stones can be directly used for masonry, bridge engineering structures, and ancillary structures. Stable rock aggregates can be used to make asphalt mixtures or cement concrete for asphalt pavement or cement pavement. They can also be used directly for paving road bases, bedding layers, or low-level road surfaces. The correct understanding, rational selection, and scientific use of aggregates are highly important for ensuring project quality and reducing production costs.

3.2.1 Classification and Sources of Aggregates

An aggregate is a mixture of mineral particles of different sizes, usually consisting of sand, gravel, slag or crushed stone, which act as a skeleton or filler in asphalt mixtures or cement concrete. It can be classified according to the source of collection or the size of the aggregates.

I. **Classification by sources of aggregates**

(a) Natural aggregates

Natural aggregates include natural sand, pebbles, or gravel.

Natural sand refers to naturally occurring, artificially mined, and sieved rock particles less than 4.75 mm in size, including river sand, lake sand, mountain sand, and desiccated sea sand, but excludes soft, weathered rock particles.

Pebble and gravel are rock particles greater than 4.75 mm in size formed by natural weathering, water transport and sorting, and accumulation. Gravel and pebble particles are usually smooth and unangular and can be further crushed for use as crushed gravel.

(b) Artificial aggregates

Artificial aggregates are mineral particles with rough angular surfaces, such as crushed stone, crushed gravel, sand, and stone chips, made from rocks, pebbles, mine waste rock, or industrial slag by crushing and screening mechanical equipment.

Crushed stone or crushed gravel consists of natural rock, pebbles, or mine waste rock that has been mechanically crushed and screened, with a particle size greater than 4.75 mm.

Manufactured sand refers to rock, mine tailings, or industrial waste particles with a particle size of less than 4.75 mm (or 2.36 mm) but does not include soft, weathered particles, commonly known as artificial sand, after soil removal, which are made by mechanical crushing and sieving. It can also be mixed with a certain proportion of mechanical sand, natural sand, or stone chips.

Chips, also known as stone chips, are the part of the quarry that passes through the smallest sieve (usually 2.36 mm or 4.75 mm) when processing crushed stone.

The process of producing artificial aggregates involves mainly the crushing and screening of rock or pebbles, or mine waste rock. The main crushing methods currently in use are crushing, impact, grinding, and splitting. The usual practice is to feed the mined rock into an impact crusher for coarse crushing and then into an impact crusher (or cone crusher) for medium crushing, where some of the rock that meets the size requirements is separated from the vibrating screen and the larger rock is then passed through the impact crusher (or cone crusher) for final crushing. The qualified product is then screened and sorted according to different size specifications.

II. Classification by size of aggregate particles

The size of the aggregate particles is expressed in terms of the particle size. According to the classification of aggregate particle size, the aggregates used in the project can be divided into coarse aggregates and fine aggregates. For different types of mixtures, the sizes of the coarse and fine aggregates are divided into different sizes.

In the asphalt mixtures, the coarse aggregates are crushed stone, crushed gravel, screened gravel, and slag with a particle size greater than 2.36 mm, whereas the fine aggregates are manufactured sand, natural sand, and stone chips with a particle size less than 2.36 mm. In cement concrete, coarse aggregate refers to crushed stone, gravel, and crushed gravel with a particle size greater than 4.75 mm, and fine aggregate refers to natural sand and manufactured sand with a particle size less than 4.75 mm.

3.2.2 Physical Properties of the Aggregates

I. Density

An aggregate is a bulk mixture of mineral particles, the volume composition of which includes, in addition to minerals and intergranular pores, the space between mineral particles, known as voids. A schematic diagram of the relationship between the volume and mass of the aggregates is shown in Fig. 3.2

In engineering, the commonly used aggregate densities include apparent density, gross bulk density, surface dry density and bulk density, where the definition of aggregate gross bulk density is the same as that of rock.

(a) Apparent density

The apparent density is the mass per unit apparent volume of the mineral entity of the dried aggregate, including closed pores, under specified conditions and is calculated via Eq. (3.1). To determine the apparent volume of the aggregate, a known mass of dry aggregate needs to be dipped in water so that its open pores are saturated with

Fig. 3.2 Schematic diagram
of the mass–volume
relationship for aggregate
composition

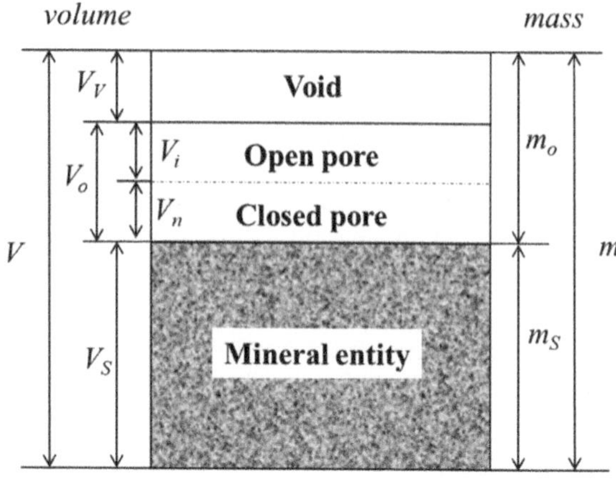

water, and then the mass of the aggregate in water after saturation is weighed out;
the difference between the two is the apparent volume of the aggregate, including
the closed pores $(V_s + V_n)$.

$$\rho_a = \frac{m_s}{V_s + V_n} \tag{3.1}$$

where

ρ_a is the apparent density of the aggregate (g/cm^3).
m_s is the mass of the mineral entity of the aggregate (g).
V_s is the volume of the mineral entity of the aggregate (cm^3).
V_n is the volume of closed pore space in the aggregate mineral entity (cm^3).

(b)   Saturated surface-dry density

The saturated surface-dry density of the aggregate is also known as the saturated
surface-dry gross bulk density; its calculated volume is the same as the gross bulk
density, but the calculated mass to the surface-dry mass (saturated surface in the
state, including the inhalation of water in the open pores) will prevail, as calculated
by Eq. (3.2).

$$\rho_s = \frac{m_a}{V_s + V_n + V_i} \tag{3.2}$$

where

ρ_s is the saturated surface-dry density of the aggregate (g/cm^3).
m_a is the surface-dry mass of the aggregate particles (g).
V_i is the volume of the open pore space in the aggregate particles (cm^3).
V_s is the volume of the mineral entity of the aggregate (cm^3).
V_n is the volume of closed pore space in the aggregate mineral entity (cm^3).

When testing the density of aggregates, the effect of the density of water at different temperatures during the test should be considered, and the density at the test temperature should be calculated. In engineering, the expressed density of the aggregate is used when designing the cement: concrete mix ratio. The relative density of the aggregates and the gross volume relative density are used in the design of the asphalt mixtures. The relative density of the aggregate is defined as the ratio of the density to the density of water at the same temperature, and the relationship between the two is shown in Eq. (3.3).

$$\rho = \gamma \cdot \rho_T \tag{3.3}$$

where

ρ is the density of the aggregate (g/cm^3).
γ is the relative density of the aggregate.
ρ_T is the density of water when test temperature is T (g/cm^3).

(c) Bulk density

The bulk density of an aggregate is the mass of the unit accumulation volume (including the void space between the aggregate particle aggregate mineral entity and its closed and open pore volume) of the oven-dried aggregate particle mineral entity, which is calculated according to Eq. (3.4).

$$\rho = \frac{m_s}{V_s + V_n + V_i + V_V} \tag{3.4}$$

where

ρ is the bulk density of the mineral aggregate (g/cm^3).
m_s is the mass of the mineral entity of the aggregate particles (g).
V_V is the volume of void space between aggregate particles (cm^3).
V_s is the volume of the mineral entity of the aggregate (cm^3).
V_n is the volume of closed pore space in the aggregate mineral entity (cm^3).
V_i is the volume of the open pore space in the aggregate particles (cm^3).

The aggregates are mixtures that have no fixed shape, the shape of which depends on the filling container, and the bulk density of which depends on the method of packing. The bulk volume of an aggregate is determined by loading a dry bulk aggregate sample into a container of a specified size, and the size of the bulk density depends on how tightly the particles are arranged, i.e., on the loading method.

Depending on the loading method, the bulk density of the aggregate includes the bulk density in the natural, compacted and pounded states. The natural density of accumulation is the density measured by filling the aggregate in a free-fall manner; the density of compaction is the density measured by loading the aggregate into the container barrel in layers, placing a round steel bar at the bottom of the container barrel and banging the barrel on the ground 25 times alternately after each layer of aggregate; the density of compaction is the density measured by loading the aggregate into the container in three layers and compaction of each layer with a pounding bar 25 times.

The natural packing density is also referred to as loose packing density, while the vibrating density and pounding density are collectively known as tight packing density. The test methods for coarse and fine aggregate bulk densities differ and should be selected according to the engineering application of the aggregate and its size.

II. Percentage of Voids in the Aggregate

The percentage of voids in the aggregate represents the volume of voids between aggregate particles as a percentage of the total volume of aggregate, reflecting the density with which the aggregate particles are packed together. The percentage of voids in the aggregate cannot be tested and is usually calculated from the density of the aggregate. In general, the void fraction of the aggregate is calculated according to Eq. (3.5).

$$n = \left(1 - \frac{\rho}{\rho_a}\right) \times 100 \tag{3.5}$$

where

n is the void ratio of the aggregate (%).
ρ_a is the apparent density of the aggregate (g/cm^3).
ρ is the bulk density or tight packing density of the aggregate (g/cm^3).

The voids in the coarse aggregate are defined as the percentage of the total volume of the sample that is outside the coarse aggregate fraction. In cement concrete, the coarse aggregate void ratio is calculated from the bulk density of coarse aggregate in the vibrated state, as shown in Eq. (3.6).

$$V_G = \left(1 - \frac{\rho}{\rho_a}\right) \times 100 \tag{3.6}$$

where

V_G is the void ratio of the coarse aggregate in the cement concrete (%).
ρ_a is the apparent density of coarse aggregate (kg/m^3).
ρ is the bulk density of the coarse aggregate (kg/m^3).

In the asphalt mix composition design, to evaluate whether the coarse aggregate used to form a skeletal structure and analyze the fine aggregate content of the mixture combined with the content of the material being reasonable, the coarse aggregate clearance rate VCA_{DRC}, which uses coarse aggregate in the pounded state of the stack density, is calculated according to Eq. (3.7). This indicator is mainly used for the design of the composition of SMA mixtures or OGFC mixtures.

$$VCA_{DRC} = \left(1 - \frac{\rho}{\rho_b}\right) \times 100 \tag{3.7}$$

where

VCA_{DRC} is the coarse aggregate interstitial ratio under pounded conditions (%).

ρ is the bulk density of coarse aggregate as determined by the pounding method (kg/m^3).

ρ_b is the bulk density of the coarse aggregate (kg/m^3).

The test results reveal that common rock aggregates are in the natural state of accumulation, with a coarse aggregate void ratio ranging from 43 to 48% and a fine aggregate void ratio ranging from 35 to 50%; in the vibration state or pounding state, the coarse aggregate void ratio ranges from 37 to 42%, and the fine aggregate void ratio ranges from 30 to 40%.

III. Gradation

Gradation refers to the proportion or distribution of various particle sizes in the aggregate, which has a significant effect on the strength, stability and ease of construction of cement-concrete and asphalt mixtures and is an important part of the mix design of cement-concrete and asphalt mixtures.

The grading of aggregates is determined by sieve testing by taking a number of aggregate samples and sieving them one by one on a standard set of sieves in order of hole size.

(a) Standard sieve

The standard sieve set is a series of sample sieves with the required shape and size specifications. The standard sieve is a square sieve with side dimensions of 70, 63, 53, and 37 mm. 5, 31.5, 26.5, 19, 16, 13.2, 9.5, 4.75, 2.36, 1.18, 0.6, 0.3, 0.15 and 0.075 mm. Owing to the different particle size ranges of the coarse and fine aggregates, the ranges of standard sieve sizes and specimen masses used in the sieving tests differ.

(b) Gradation parameters

In the sieve test, the sieve residual mass of the aggregate sample is weighed on each sieve, and then the relevant parameters reflecting the gradation of the aggregate sample are calculated.

The percentage of sieve residue, a_i, is the percentage of the total mass of the sample on a particular sieve and is calculated according to Eq. (3.8).

$$a_i = \frac{m_i}{m} \times 100 \tag{3.8}$$

where

m_i is the mass of aggregates deposited on a sieve (g).

m represents the mass of the dried aggregates (g).

The cumulative percentage of sieve residue, A_i, is the sum of the percentage of sieve residue counted for a particular sieve and the percentage of sieve residue counted for each sieve larger than that sieve and can be found in Eq. (3.9) as follows:

$$A_i = a_1 + a_2 + \cdots + a_i \tag{3.9}$$

where, $a_1, a_2, \ldots, a_i$ represents the residual percentage of each sieve (%).

The passage percentage p_i is the percentage of the mass of the sample passing through a particular sieve to the total mass of the sample, i.e., the difference between 100 and the cumulative sieve residual percentage of a particular sieve, obtained according to Eq. (3.10).

$$p_i = 100 - A_i \tag{3.10}$$

where, A_i is the cumulative percentage of sieve residue of a particular sieve (%).

(c)   Drawing of the gradation curve

The sieving test results of the aggregate are expressed in terms of the percentage of mass passing through each sieve and can also be expressed in terms of a gradation curve. In the gradation curve diagram, the percentage passage (or cumulative sieve margin) is usually indicated by the vertical coordinate, and the horizontal coordinate indicates the sieve size of a particular sieve, as shown in Fig. 3.3.

In a standard set of sieves, the sieve hole size decreases by approximately 1/2, and if both the vertical and horizontal coordinates of the grading curve are expressed in constant coordinates, the sieve hole size on the horizontal coordinate is dense on the left and sparse on the right, as shown in Fig. 3.3a. To facilitate drawing and access, the horizontal coordinates are usually logarithmic so that most sieve hole sizes are arranged at equal intervals on the horizontal coordinates, as shown in Fig. 3.3b. When the grading curve is drawn, first, the logarithmic coordinate position of the sieve hole size on the horizontal coordinate is marked, the constant coordinate

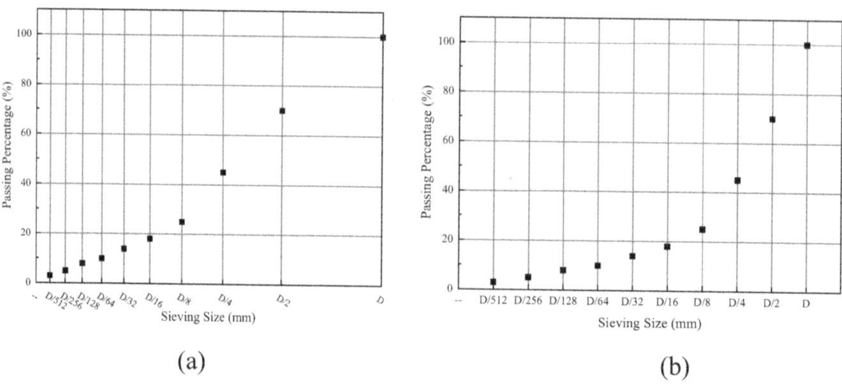

Fig. 3.3 Aggregate grading curve diagram

position of the through percentage (or cumulative sieve residual percentage) on the vertical coordinate is marked, the result points of the sieve test on the coordinate chart are plotted, and finally, the points are connected into a grading curve. More than two gradation curves can be plotted simultaneously on the same graph, provided that the aggregate species represented by each curve are indicated.

(d) Calculation of the fineness module

The fineness module is an index for evaluating the coarseness of natural sand and is the sum of the cumulative percentage of sieve residue on each number of sieves in the natural sand sieving test, which is calculated according to Eq. (3.11). The higher the fineness modulus is, the coarser the sand. Sand is divided into coarse sand (3.7 –3.1), medium sand (3.0–2.3), and fine sand (2.2–1.6) according to the fineness modulus.

$$M_f = \frac{(A_{2.36} + A_{1.18} + A_{0.60} + A_{0.30} + A_{0.15}) - 5A_{4.75}}{100 - A_{4.75}} \tag{3.11}$$

where

M_f	is the fineness module of the sand.
$A_{4.75}, A_{2.36},...., A_{0.15}$	are the cumulative percentages of sieve residue for sieves of 4.75 mm, 2.36 mm,..., 0.15 mm (%), respectively.

(e) Effect of the graded composition on the mineral properties

The shape of the grading curve of the mineral mixture can be divided into continuous gradation and gap gradation. In the continuous gradation type of ore, from large to small and all levels of particle size, all levels of particles are in accordance with a certain ratio, and the grading curve is drawn smoothly and continuous without interruption. In the intermittent gradation type of ore, there is a lack of particles of one or several grades, there is a large gap between the large and small particles, and the grading curve is drawn in a discontinuous, interrupted shape.

The grade composition of a mineral mixture is closely related to its compactness and the internal frictional resistance between particles, thus having a significant effect on the strength, durability, construction and ease of cement-concrete or asphalt mixing. Generally, the void ratio of continuously graded minerals increases significantly with increasing coarse aggregate content, whereas intermittently graded minerals can better serve as the skeleton of coarse aggregates but can easily segregate during the construction process.

In cement concrete or asphalt mixtures, the binding material (cement or asphalt) fills the aggregate voids and encapsulates the aggregate. Therefore, the larger the aggregate void is, the more binding material is required to fill the void of the aggregate particles; the larger the total surface area of the aggregate is, the more binding material is required to wrap the aggregate particles. From the point of view of saving binding material, it is best to use aggregates with smaller voids and a smaller total surface area. In addition, if the granular aggregate particles in a mutual arrangement can interlock and do not interfere with each other, forming a tight multilevel embedded space skeleton structure, then the aggregate particles will have greater internal friction resistance.

IV. Particle shape and surface characteristics

The particle shape and surface characteristics of aggregates, especially coarse aggregates, significantly influence the internal frictional resistance between aggregate particles and the adhesion of aggregate particles to the bonding material at the interface.

(a) Particle shape

From a practical point of view, the shape of aggregate particles can be divided into four types according to Table 3.1, and a more ideal shape is close to a sphere or cube. When the content of flat, flaky, elongated particles in the aggregate is high, the void ratio of the aggregate increases, which not only impairs the construction and ease of the aggregate but also jeopardizes the strength of the concrete to varying degrees.

(b) Surface features

The surface features of the aggregate mainly refer to the roughness and pore characteristics of the aggregate surface, which are related to the material of the aggregate, the rock structure, the mineral composition and the extent to which it is scoured and corroded. In general, the surface characteristics of the aggregate mainly affect the bonding properties between the aggregate and the bonding material, thus affecting the strength of the mixture, especially the flexural strength. Under the action of external forces, the displacement between particles of aggregates with rough surfaces is more difficult, and their frictional resistance is greater than that of smooth, unangled particles, but this affects the construction and ease of aggregation. In addition, aggregates with rough surfaces and pore characteristics that absorb the lighter components of cement paste or asphalt have a stronger bonding ability to the bonding material, whereas aggregates with smooth surfaces generally have a poorer bonding ability to the bonding material.

The surface characteristics of fine aggregates such as natural sand, artificial sand, and stone chips significantly influence the internal friction angle and resistance to mobility deformation of asphalt mixtures and the compatibility of cement concrete.

Table 3.1 Basic types of aggregate particle shapes

Type	Aggregate particle shapes	Aggregate species
Egg round	Smooth surface, no obvious angles, rounded grains	Natural sand and various gravels, ceramic granules
Angular	A rough surface and pronounced edges	Crushed stone, stone chips, crushed slag
Needle-shaped	Lengthwise dimensions are much larger than other directions and are in the shape of thin strips	Present in both gravel and crushed stone
Flake	The thickness direction is much smaller than the other directions and has a thin sheet shape	Present in both gravel and crushed stone

The surface characteristics of fine aggregates are characterized by the angularity index, which can be assessed via the void ratio method or the flow time method.

V. **Mud and clay contents**

The presence of clay in the aggregate or wrapped around the surface of the aggregate particles can reduce the rate of hydration of the cement and hinder the bonding ability between the aggregate and the cement or asphalt, significantly affecting the overall strength and durability of the mixture; thus, its content should be limited.

(a) Clay content and fine content

The clay content is the amount of particles less than 0.075 mm in size in natural sand, gravel or pebbles; the fine content is the amount of particles less than 0.075 mm in size in manufactured sand. Both can be calculated according to Eq. (3.12).

$$Q_a = \frac{m_0 - m_1}{m_0} \times 100 \tag{3.12}$$

where

Q_a is the mud or stone dust content (%).
m_0 is the mass of the dried sample before the test (g).
m_1 is the mass of the dried sample on a 0.075 mm sieve after sieving (g).

(b) Clay lump and friable particle contents

The clay lump content is the content of particles in pebbles and aggregates with an original particle size greater than 4.75 mm (1.18 mm in sand) but less than 2.36 mm (0.6 mm in sand) after washing with water and hand kneading, which is calculated according to Eq. (3.13). Mud clumps in aggregates are present in three main types: clumps consisting of pure clay; clumps consisting of sand, stone chips and clay; and mud encapsulated on the surface of aggregate particles.

$$Q_b = \frac{m_2 - m_3}{m_2} \times 100 \tag{3.13}$$

where

Q_a is the mud or stone dust content (%).
m_0 is the mass of the dried sample before the test (g).
m_1 is the mass of the dried sample on a 0.075 mm sieve after sieving (g).

3.2.3 Mechanical Properties of the Aggregates

In the mix, coarse aggregates play a skeletal role and should have a certain strength, wear resistance, abrasion resistance, impact resistance, etc. These properties, including the crushing value, grinding value, abrasion value, impact value and other indicators, are expressed.

I. Crush Stone Value

The crush stone value is used to measure the ability of an aggregate to resist crushing under a gradually increasing load and is also a relative indicator of aggregate strength; it is used to identify the quality of the aggregate and determine its suitability for use in road engineering.

In the crushing value test, an aggregate sample with a particle size of 9.5–13.2 mm is loaded into a mold, uniformly loaded to 400 kN in approximately 10 min and unloaded after 5 s of steady pressure. The samples were removed and sieved with a 2.36 mm standard sieve, and the masses of all the fines passing through the 2.36 mm sieve were then weighed. The crushing value of the coarse aggregate, Q_a', is calculated according to Eq. (3.14).

$$Q_a' = \frac{m_1}{m_0} \times 100 \tag{3.14}$$

where

Q_a is the crush stone value of the coarse aggregates (%).
m_0 is the mass of the coarse aggregates before the test (g).
m_1 is represents the mass of coarse aggregates on a 2.36 mm sieve after sieving (g).

The crushing index of the concrete sand is determined by dividing the sand into four grades: 2.36–4.75 mm, 1.18–2.36 mm, 0.6–1.18 mm and 0.3–0.6 mm, and loading the samples of each grade under standard conditions. After loading, the samples are sieved with a lower limit sieve for each grade (e.g., 2.36–4.75 mm samples are sieved with a 2.36 mm standard sieve), the sieve residual and throughput of each grade are weighed, and the crushing index value for each grade is calculated according to Eq. (3.15). The maximum single-grain crushing index value is taken as the crushing index value of the sand.

$$Y_i = \frac{m_1}{m_1 + m_2} \times 100 \tag{3.15}$$

where

Y_i is the crushing index value for the ith grain of the fine aggregates (%).
m_0 is the sieve residual of the fine aggregates (%).
m_1 is the passing percentage of fine aggregates (%).

II. Abrasion

Abrasion is one of the most important indicators for evaluating the performance of aggregates and is closely related to the rutting resistance, wear resistance and durability of asphalt pavements. Abrasion loss is used to assess the ability of coarse aggregates to resist friction and impact, and the abrasion value is used to assess the ability of coarse aggregates used on road pavement surfaces to resist the abrasive effect of wheels.

(a) Abrasion Loss

The abrasion loss test is carried out via the Los Angeles abrasion tester. First, according to the composition of the grain level of the aggregate, according to the provisions of the preparation of the samples and steel balls, a certain grade of aggregate samples and steel balls were placed in a abrasion test machine, the abrasive machine was started, the rotation speed was 30–33 r/min to the required number of rotations, and then the process was stopped. The steel balls were removed, the sample was sieved, washed, dried and weighed. The abrasion loss of the aggregate is calculated according to Eq. (3.16).

$$Q = \frac{m_1 - m_2}{m_1} \times 100 \qquad (3.16)$$

where

Q is the abrasion loss (%).
m_0 is the mass of the aggregate before the test (g).
m_1 represents the mass of the aggregate on a 1.7 mm sieve after sieving (g).

(b) Weared stone value

The weared stone value test is carried out via a Dorry abrasion tester. Aggregates of 9.5–13.2 mm are arranged in a single layer in a tightly packed test mold with no fewer than 24 particles, filled with epoxy resin mortar and demolded after curing to form the sample. Two samples of the same aggregate were fixed to the circular plate of the Dawry abrasion tester and rotated at 28–30 r/min for 100 revolutions while the specified fineness of the quartz sand was continuously spread onto the grinding disc. After the machine was stopped, the test pieces were removed, and any abnormalities were observed. Then, 400 revolutions were ground according to the same method, which can be divided into 100 revolutions and repeated 4 times to finish the grinding or 1 time continuously to finish the grinding. After the machine was stopped, the mass of the test pieces was weighed, and the abrasion value of the aggregate was calculated according to Eq. (3.17).

$$AAV = \frac{3(m_1 - m_2)}{\rho_S} \qquad (3.17)$$

where

AAV is the Dorry weight value for aggregates (%).
m_1 is the mass of the aggregate before the test (g).
m_2 is the mass of the aggregate after the test (g).
ρ_S is the surface dry density of the aggregates (g/cm^3).

III. Polished Stone Value

The polished stone value is an indicator of the ability of an aggregate to resist the abrasive effects of tyres. The use of aggregates with a high polished stone value can improve the skid resistance of the road surface and ensure safe vehicle operation.

The basic method involves placing 9.5–13.2 mm clean aggregate particles in a single layer in a tightly packed test mold and fixing them with an epoxy resin mortar to make a test piece. Two samples of the same aggregate were mounted on the road wheel in sequence with other aggregate samples and standard aggregate samples, first with 30 grit particles for 3 h and then with 280 grit particles for 3 h before stopping. After the samples were removed, a pendulum friction coefficient tester was used to determine the abrasive value of the samples (friction coefficient), and the abrasive value of the aggregate was calculated via Eq. (3.18).

$$PSV = PSV_{ra} + 0.49 - PSV_{br} \tag{3.18}$$

where

PSV is the polished stone value for the aggregates (British Pendulum Number, BPN).

PSV_{ra} is the average abrasion value of the aggregates (BPN).

PSV_{br} is the average abrasion value of the standard aggregates (BPNs).

IV. Impact Value

The impact value reflects the ability of the coarse aggregate to resist impact loading. As the road surface aggregate is directly impacted by the wheel load, this indicator is very important for road surface aggregates.

The impact value of the aggregate is tested using a dry aggregate of size 9.5–13.2 mm, which is loaded into a measuring cylinder in three layers according to standard methods, and the mass of the aggregate sample is weighed. The weighed aggregate is loaded into a circular steel cylinder, placed on a impact testers and pounded individually with a tamping rod 25 times. The hammer height was adjusted so that the hammer fell freely from 380 to 5 mm, and the aggregate was hammered 15 times continuously at intervals of no less than 1 s. The aggregates were sieved through a 2.36 mm sieve after the impact test, and the mass of the stone chips passing through the 2.36 mm sieve was determined. The impact value of the aggregate is calculated according to Eq. (3.19).

$$AIV = \frac{m_1}{m} \times 100 \tag{3.19}$$

where

AIV is the impact value of the aggregates (%).

m is the total mass of the sample (g).

m_1 is the mass of the aggregate on a 2.36 mm sieve (g).

Summary

Aggregates are the most commonly used type of material in road engineering and can be used directly for paving road structures; however, they are more often prepared as asphalt mixtures, cement concrete and base mixes for paving road surfaces, base layers or bedding layers.

In road engineering, the commonly used rock types are limestone, granite, sandstone, basalt and gabbro. The main physical constants of the rock are density, water content and water absorption; the main mechanical index is the uniaxial unconfined compressive strength, and the frost resistance of the rock used should be considered in seasonal freezing areas.

An aggregate is a mixture of mineral particles of different sizes. The density and void ratio of aggregates have important influences on their physical and mechanical properties. The mechanical properties of the aggregates are expressed in terms of the crushing value, abrasion loss and wear value, impact value and abrasion value. The coarse aggregates used in road pavement structures should have sufficient crush resistance, abrasion resistance and impact resistance, and the coarse aggregates used in asphalt pavement surfaces should also have sufficient abrasion resistance.

The gradation composition of the aggregate reflects the distribution of the different particle sizes and is expressed in terms of gradation parameters and gradation curves. It has a significant influence on the compactness and internal friction resistance of the aggregate and is the main basis for the design of the mineral mix composition. The gradation types of mineral materials are divided into continuous gradation and intermittent gradation; the former can be calculated via the gradation curve range formula.

3.3 Asphalt and Asphalt Mixtures

3.3.1 Asphalt Material

Classification of Asphalt

Asphalt is a black or dark black solid, semisolid or viscous substance made by natural or artificial manufacturing and is composed mainly of high-molecular-weight hydrocarbons, which can be dissolved in carbon disulfide. Its chemical composition and microstructure are extremely complex. As a common binder in road materials, asphalt is used to bind loose aggregate particles into a whole material with a certain strength and stability.

There are many varieties of asphalt. The broad sense of asphalt mainly includes natural asphalt, tar asphalt and petroleum asphalt, whereas the narrow sense of asphalt mainly refers to petroleum asphalt. Petroleum asphalt is a chemical product refined from crude oil by a specific production and processing technology, and it is the most widely used asphalt material at present.

Petroleum asphalt is divided into distilled asphalt, oxidized asphalt, solvent asphalt and blended asphalt according to processing methods. The state of asphalt at room temperature can be divided into viscous asphalt and liquid asphalt. Viscous asphalt, which is a relatively high viscosity asphalt, is paste-like or solid at room temperature. Liquid asphalt refers to asphalt that is in a liquid or semifluid state at room temperature. The dilute asphalt was obtained by diluting viscous asphalt with solvent. Emulsified asphalt is another form of liquid asphalt. Emulsified asphalt can be divided into cationic emulsified asphalt, anionic emulsified asphalt and nonionic emulsified asphalt according to the type of emulsifier used.

Technical Properties of Petroleum Asphalt

The physical properties of petroleum asphalt can be characterized by several physical constants, such as density, the volume expansion coefficient and the dielectric constant. The density of asphalt is the mass per unit volume at a specified temperature. The relative density of asphalt refers to the ratio of the mass of asphalt to the mass of water with the same volume at the specified temperature. The relative density of asphalt is closely related to the chemical composition of the asphalt. It depends on the proportion of the asphalt components and the tightness of the arrangement. Asphalt with a high sulfur content, high aromatic content and high asphaltene content has a relatively high relative density. The higher the wax content is, the lower the relative density. The volume expansion coefficient of asphalt varies in the range of 2×10^{-4} to $6 \times 10^{-4}/°C$. The volume expansion coefficient of asphalt has a certain relationship with road performance. The larger the expansion coefficient is, the easier it is for the asphalt pavement to flood oil in summer, and the more prone it is to shrinkage and cracking in winter. The dielectric constant of asphalt is related to the weather resistance (aging resistance) to oxygen, rain and ultraviolet rays. According to the research of the TRL (Transport Research Laboratory), formerly the British government's Road Research Laboratory (RRL), the improvement in asphalt pavement skid resistance is related to the dielectric constant, so the British standard puts forward requirements for the dielectric constant of asphalt.

The road properties of asphalt include viscosity, low-temperature performance, temperature sensitivity, adhesion, durability and viscoelasticity.

(1) Viscosity

Viscosity refers to the ability of an asphalt material to resist shear deformation under an external force. Asphalt should be able to bond loose mineral materials into a whole. Therefore, viscosity is the most important property of asphalt. The dynamic viscosity calculation formula of asphalt is shown in Eq. 3.20.

$$\eta = \frac{\tau}{\dot{\gamma}^c} \tag{3.20}$$

where

η is the dynamic viscosity of the asphalt (Pa)· s).
τ is the shear stress (N/m^2).
$\dot{\gamma}$ is the shear rate (s^{-1}).
c   is the composite flow coefficient of asphalt.

The composite flow coefficient c value reflects the rheological properties of the asphalt. Under high-temperature conditions, asphalt exhibits Newtonian liquid viscosity; that is, the relationship between the shear stress and shear rate is linear, and the value of c is close to 1. However, in the service temperature range of pavement, asphalt is viscosity-elastic–plastic, and its shear stress has a nonlinear relationship with the shear rate, with a value of c generally less than 1.

The viscosity of the moving state is expressed by the kinematic viscosity, and the kinematic viscosity is expressed by Eqs. 3.21.

$$v = \frac{\eta}{\rho} \tag{3.21}$$

where

v is the kinematic viscosity of the asphalt (mm^2/s).
η is the dynamic viscosity of the asphalt (Pa)· s).
ρ is the density of the asphalt (g/cm^3).

The viscosity of asphalt varies with temperature, and different instruments and methods are needed for measurement. To measure the viscosity grade of the asphalt at 60 °C, the dynamic viscosity of the asphalt is measured via a vacuum decompression capillary viscometer. At a construction temperature of 135 °C, the kinematic viscosity is usually measured via the capillary method. The apparent viscosity and shear viscosity can be measured by a Brookfield rotary viscometer and a dynamic shear rheometer, respectively. These methods for measuring viscosity are called the absolute viscosity method. In addition, several empirical methods can be used to measure the relative viscosity, such as the Engler viscometer and the road asphalt standard viscometer. In addition, the penetration test and softening point test can also characterize the relative viscosity of asphalt.

(2)   Low-temperature performance

The low-temperature performance of asphalt is closely related to the low-temperature crack resistance of asphalt pavement. Ductility tests, brittleness tests, bending beam rheometer tests and direct tensile tests are commonly used to measure the low-temperature performance of asphalt.

The ductility of asphalt refers to the total capacity of plastic deformation that it can withstand when it is stretched by external forces, and it is a measure of the cohesion of asphalt. The ductility is closely related to the rheological properties, colloidal structure and chemical composition of the asphalt. Research has shown that the chemical composition of asphalt is uncoordinated, the colloid structure is uneven, and the wax content increases, all of which decrease the ductility value of asphalt.

When asphalt is subjected to an instantaneous load at low temperature, it shows brittle failure. Usually, the A. Fraser brittle point test can be used to calculate the temperature when the asphalt reaches the critical hardness and cracks, which can be used as the brittleness index. The brittle point essentially reflects the temperature at which the asphalt changes from viscoelastic to elastic brittle and represents the temperature at which the asphalt reaches equal stiffness. The stiffness of asphalt when brittle cracks occur is approximately 2.1×10^9 Pa.

The results of studies have shown that the low-temperature stiffness of an asphalt mixture is an important indicator reflecting its anti-cracking performance[1, 2]. SHRP is a method that can accurately evaluate the stiffness and creep rate of asphalt, that is, the bending beam rheometer (BBR) test. The BBR test can predict and evaluate the low-temperature performance of asphalt pavement after five years of service. The low-temperature performance evaluation indices are the creep stiffness modulus and the slope of the creep curve. If the creep stiffness of asphalt is too high, it will be brittle, and the pavement will easily crack. Therefore, the creep stiffness modulus must not exceed 300 MPa [3]. However, the greater the rate of change in the asphalt stiffness at low temperatures with time is, the lower the possibility of asphalt cracking.

Direct tensile tests were developed by SHRP to test the tensile properties of asphalt and are used to test the ultimate tensile strain of asphalt at low temperatures. The SHRP specification requires that the failure strain of the direct tensile test should not exceed 1% [4].

(3) Temperature sensitivity

Asphalt viscosity changes significantly with temperature, and the sensitivity of this viscosity to temperature changes is called temperature sensitivity. The relationship between temperature and viscosity is extremely important. First, the temperature sensitivity of asphalt significantly reduces the viscosity at high temperatures so that it is possible to achieve uniform mixing and compacting of the asphalt mixture. Second, during the service life of asphalt pavement, asphalt is required to maintain a low temperature sensitivity to ensure that the asphalt pavement does not soften at high temperatures and does not crack at low temperatures. Commonly used methods to evaluate the temperature sensitivity of asphalt include the penetration index (PI) method and the penetration-viscosity index (PVN) method. In addition, the softening point test can also be used as a method to reflect the temperature sensitivity of asphalt.

(4) Adhesion

In an asphalt mixture, asphalt is coated on the surface of an aggregate in the form of a film. In addition to the viscosity of the asphalt itself, it also needs adhesion between the asphalt and the stone. In general, asphalt with higher viscosity has greater adhesion. The adhesion of asphalt is characterized by boiling and water immersion methods. Under dry conditions, asphalt generally has sufficient adhesive strength. However, moisture is one of the reasons for adhesion problems. In addition, owing to the repeated action of traffic loads, the pavement is deformed, and the aggregate is loose. The asphalt film and the aggregate are peeled off by immersion, resulting in the destruction of the asphalt pavement.

(5) Durability

Asphalt is subjected to storage, transportation, heating, mixing, paving, rolling, traffic loading and natural factors in the process of use, which causes it to undergo a series of physical and chemical changes, gradually changing its original properties (viscosity, low-temperature performance). This change is called the aging of asphalt. Asphalt pavement should have a long service life, and asphalt materials must have good aging resistance. The current test methods for evaluating asphalt aging performance are divided into simulated thermal oxygen aging conditions during mixing and aging conditions during service, such as the film oven heating test, rotating film heating test, pressure aging vessel test, and weathering aging test.

(6) Viscoelasticity

Asphalt is elastic at low temperatures, viscous at high temperatures, and both viscous and elastic over a wide temperature range. It is a typical viscoelastic material. Deformation occurs under the action of an external force, but the deformation lags behind the force. After the force is removed, the deformation is not completely eliminated, and the deformation gradually recovers after a period of time, resulting in complex viscoelastic properties, such as creep and relaxation.

Under traffic loading, the relationship between stress and strain is nonlinear. Therefore, the stiffness modulus is used to describe the mechanical properties of asphalt in the viscoelastic state. The ratio of stress to strain is used to express the deformation resistance of viscoelastic asphalt, and the stiffness modulus is expressed by Eqs. 3.22. The stiffness modulus of asphalt can be measured by instruments such as a microfilm slide viscometer or a micro elasticity meter, and can also be determined via a nomogram of the asphalt stiffness modulus.

$$S = \frac{\delta}{\varepsilon_{t,T}} \tag{3.22}$$

where

S   is the stiffness modulus of the asphalt (Pa).
δ is the stress (Pa).
ε is the strain.
t   is the load time (s).
T   is the temperature (°C).

The SHP program uses a dynamic shear rheometer to evaluate the high-temperature stability of asphalt and measures the complex shear modulus (G^*) and phase angle (δ) to characterize the viscosity and elasticity properties of the asphalt. The complex shear modulus is a measure of the total resistance of materials, as it undergoes repeated shear deformation, including elastic (recoverable) and viscous (unrecoverable) deformation. The phase angle is a relative indicator of recoverable deformation versus nonrecoverable deformation. In most cases, asphalt is viscous and elastic.

The Federal Highway Administration (FHWA) proposed a multistress creep recovery test, which characterized the viscoelastic response and resistance to permanent deformation via the strain recovery rate and nonrecoverable creep compliance. The multistress creep recovery test is a repeated loading and unloading process consisting of 20 cycles, each of which lasts for 10 s. At the beginning of each cycle, the sample is loaded with a specified shear stress for 1 s and then unloaded for 9 s. The loading stress of the first 10 cycles is 100 Pa, and the shear stress of the last 10 cycles is 3200 Pa [5].

Modified Asphalt

Modified asphalt refers to asphalt binder made by adding external additives (modifiers), such as rubber, resin polymers, rubber powder or other fillers, or taking measures such as mild oxidation of asphalt to improve the performance of asphalt or asphalt mixtures. Broadly speaking, any material that can improve the road performance of asphalt, such as polymers, fibers, anti-stripping agents, rock asphalt and fillers (such as sulfur and carbon black), can be used as modifiers. In general, most modified additives can improve the high-temperature performance of asphalt, but the improvements in the low-temperature cracking resistance, water damage resistance and fatigue cracking performance differ.

The polymer-modified asphalts used in road engineering mainly include the following four categories:

(1) Thermoplastic rubber-modified asphalt. Thermoplastic elastomers are elastic materials produced by thermoplastic plasticization of rubber elastomers and thermoplastic technology of elastomers and resins. Among them, SBSs are widely used. The SBS polymer chain has different blocks with a block structure, namely, a plastic segment and a rubber segment. The structure of a polymer can be divided into linear SBS and radial SBS. The modification effect of radial SBS on asphalt is better than that of linear SBS. The larger the molecular weight of SBS is, the more obvious the modification effect, but processing is more difficult. Thermoplastic elastomer-modified asphalt has good temperature stability, which obviously improves the high- and low-temperature performance of matrix asphalt, reduces the temperature sensitivity, and enhances the aging resistance and fatigue resistance.

(2) Rubber-based modified asphalt. Styrene-butadiene rubber (SBR) and chloroprene rubber (CR) are the most widely used rubber-modified materials. Styrene-butadiene rubber is an early developed asphalt modifier. This kind of modifier is often added to asphalt in the form of latex, which can improve the viscosity, toughness and softening point; reduce the brittle point; and improve the ductility and temperature sensitivity of the asphalt. This is due to the swelling caused by rubber absorbing the oil in the asphalt, which changes the colloidal structure of the asphalt, thus improving the colloidal structure and viscosity of the asphalt.

(3) Thermoplastic resin-modified asphalt. Thermoplastic resin is a type of polyolefin polymer, and the commonly used varieties are low-density polyethylene (LDPE), ethylene–vinyl acetate copolymer (EVA), and APAO. Resin-modified asphalt has good high-temperature stability and rutting resistance, but it does not obviously improve the low-temperature cracking resistance of asphalt pavement.

(4) Thermosetting resins include polyurethane (PU), epoxy resin (EP) and unsaturated polyester resin (UP), among which epoxy resin has been applied to modify asphalt. Epoxy resin is an oligomer or polymer containing two or more epoxy or epoxy groups of ether or phenol. Epoxy resin-modified asphalt has poor extensibility, but it has high strength, excellent permanent deformation resistance, and particularly high resistance to fuel oil and lubricating oil. It is suitable for bus stops and gas station pavements.

Modified asphalt has different technical characteristics. In addition to the penetration, softening point, ductility and viscosity of conventional asphalt tests, several technical indices different from those used to evaluate the performance of modified asphalt are adopted, such as the polymer-modified asphalt segregation test, asphalt elastic recovery test, viscosity toughness test and force ductility test.

3.3.2 Asphalt Mixture

Classification of the Asphalt Mixture

An asphalt mixture is a mixture of mineral aggregates and asphalt. The classification method of an asphalt mixture depends on the graded composition of the mineral aggregates, nominal maximum particle size of the aggregate, compaction void ratio of the asphalt mixture, asphalt variety and preparation method of the asphalt mixture.

(1) Classification according to the aggregate gradation type

According to the characteristics of the mineral aggregate graded composition and the residual void ratio of the asphalt mixture after compaction, the asphalt mixture is classified as follows:

1. Dense-graded asphalt mixture

The principles of the continuous dense graded design composition of minerals and asphalt are as follows: dense graded asphalt concrete mixtures with a designed void ratio of 3–6%, which are usually referred to as asphalt concrete mixtures (AC), and dense graded asphalt stabilized gravel mixtures with a designed void ratio of 3–6%, which are referred to as asphalt treated bases (ATB).

2. Semi open-graded asphalt mixture

A Semi-open-graded asphalt mixture is composed of an appropriate proportion of coarse aggregate, fine aggregate and a small amount of filler (or no filler) mixed with the asphalt. The typical type is a half-open asphalt-stabilized gravel mixture with a designed void ratio of 6–12%, which is expressed as an asphalt-treated mixture (AM).

3. Open-graded asphalt mixture

The mineral gradation is mainly composed of coarse aggregate, less fine aggregate and filler, and the mixture is mixed with high viscosity asphalt binder. Its typical type is a drainage asphalt friction course with a design void ratio of 18–25%, which is expressed as an open-graded friction course (OGFC), and a drainage asphalt-stabilized gravel mixture with a design void ratio greater than 18%, which is expressed as an asphalt-treated permeable base (ATPB).

4. Gap-graded asphalt mixture

Asphalt mixtures with gap grades are formed by the lack of one or several particle size grades in the mineral aggregate graded composition. The typical type is stone matrix asphalt (SMA). The SMA is an asphalt mixture composed of asphalt binder, a small amount of fiber stabilizer, fine aggregate and more filler (mineral powder), which fills the gaps in the coarse aggregate skeleton with gap gradation (Fig. 3.4).

(2) Classification according to the nominal maximum particle size of the aggregates

The asphalt mixtures are divided into extracoarse, coarse, medium, fine and sandy mixtures according to the nominal maximum particle size of the aggregate. The types of asphalt mixtures are summarized in Table 3.2.

(3) Classification according to the mixing temperature of the asphalt mixture

1. Hot mix asphalt

The term "hot mix asphalt" refers to heating asphalt to 150–170 °C and mineral aggregate to 170–190 °C, mixing and paving construction in the hot state. The hot mix asphalt has high strength and excellent road performance, which is suitable for all layers of high-grade road asphalt pavement structures.

2. Cold mix asphalt

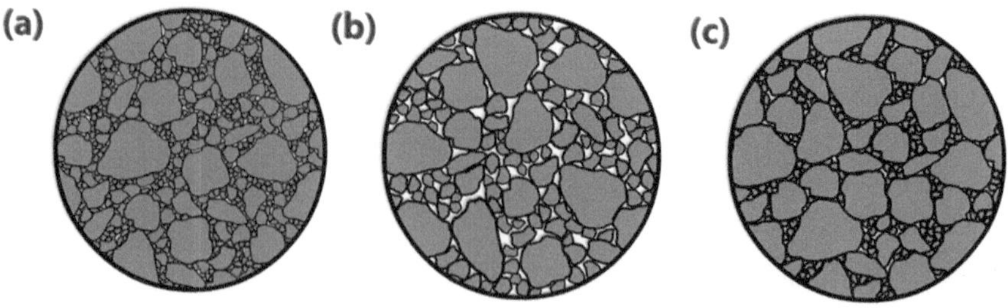

Fig. 3.4 Schematic of the asphalt mixture: **a** dense-graded asphalt mixture; **b** open-graded asphalt mixture; **c** gap-graded asphalt mixture

Table 3.2 Summary of the asphalt mixture types

Type of asphalt mixture	Nominal maximum particle size (mm)	Maximum particle size (mm)	Dense-graded		Half open-graded	Open-graded		Gap-graded
			Asphalt concrete mixture	Asphalt treated base	Asphalt-treated mixture	Open-graded friction course	Asphalt treated permeable base	Stone Matrix Asphalt
sand	4.75	9.5	AC-5	–	AM-5	–	–	–
fine	9.5	13.2	AC-10	–	AM-10	OGFC-10	–	SMA-10
	13.2	16	AC-13	–	AM-13	OGFC-13	–	SMA-13
medium	16	19	AC-16	–	AM-16	OGFC-16	–	SMA-16
	19	26.5	AC-20	–	AM-20	–	–	SMA-20
coarse	26.5	31.5	AC-25	ATB-25	–	–	ATPB-20	–
	31.5	37.5	–	ATB-30	–	–	ATPB-30	–
extra coarse	37.5	53	–	ATB-40	–	–	ATPB-40	—
Design void ratio (%)			3–6	3–6	6–12	>18	>18	3–4

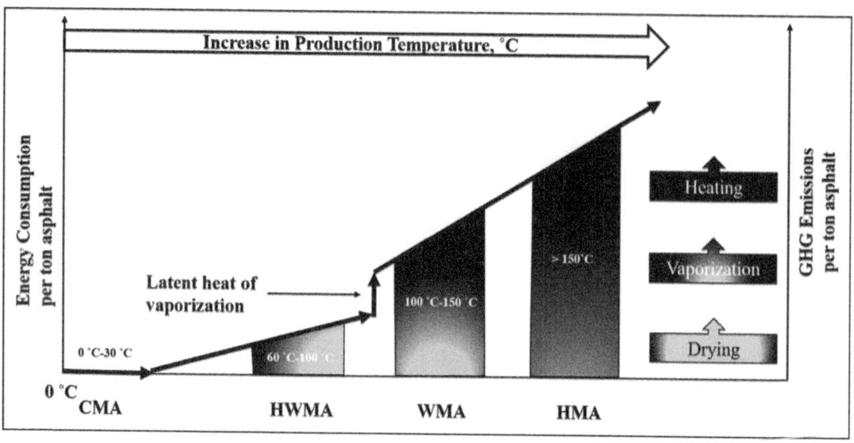

Fig. 3.5 Classification of asphalt mixtures [6]

Cold mix asphalt is also known as a normal temperature mixture. It uses emulsified asphalt, foamed asphalt, liquid asphalt or low viscosity asphalt as a binder, and the aggregate is mixed at room temperature. Then, the cold mix asphalt is paved and rolled at room temperature. Owing to the low viscosity of the asphalt used, the pavement molding time is long, and the strength is not high; thus, it is mainly used for low-grade roads and pavement repair.

3. Warm mix asphalt

Warm mix asphalt is a general name for asphalt mixtures that uses specific technologies or additives to determine the mixing, paving and compaction temperatures of asphalt mixtures between hot mix asphalt and cold mix asphalt. This is a new type of asphalt mixture production technology with energy-saving and environmental protection effects. This approach can reduce the construction temperature of asphalt mixtures, reduce harmful gas emissions, and ensure that asphalt mixtures have essentially the same road performance and workability as hot mix asphalt (Fig. 3.5).

Technical Properties of the Asphalt Mixture

When asphalt mixtures are used as surface materials of asphalt pavement, asphalt mixtures withstand repeated vehicle loads and environmental factors during the service period. Therefore, in addition to certain strengths, asphalt mixtures should also have sufficient technical properties, such as high-temperature stability, low-temperature crack resistance, water stability, aging resistance, and skid resistance, to ensure the excellent service performance and durability of asphalt pavement.

(1) High-temperature stability of asphalt mixtures

High-temperature stability refers to the ability of an asphalt mixture to resist the repeated action of vehicle loads under high-temperature conditions, prevent significant permanent deformation, and ensure the roughness of the pavement. An asphalt mixture is a typical viscoelastic–plastic material that results in obvious deformation under high temperatures or long-term loads, and the unrecoverable part becomes permanently deformed. This characteristic is the main reason for rutting, waves, and the upheaval of asphalt pavement. Rutting is one of the most serious and harmful forms of damage to asphalt pavement with heavy traffic, a high proportion of heavy vehicles and a frequent shifting speed.

There are many methods used to evaluate the high-temperature stability of asphalt mixtures, such as uniaxial (triaxial, shear) static loading, dynamic loading, repeated loading tests and repeated rolling simulation tests, such as rutting tests. In addition, the Marshall stability test can be used to evaluate the high-temperature performance of asphalt mixtures.

The rutting test is a test method that simulates rutting by vehicle tires rolling on the pavement. The test results are intuitive and have a good correlation with the rutting depth of asphalt pavement. The rutting test uses the standard method to shape the asphalt mixture plate sample. Under the specified temperature conditions, the test wheel repeatedly walks along the same track on the sample surface at a frequency of 42 times/min $\pm$ 1 time/min, and the rutting depth of the sample surface under repeated action is tested. Dynamic stability is used as the evaluation index of the rutting test. The calculation formula is shown in Eqs. 3.23.

$$DS = \frac{(t_2 - t_1) \cdot N}{d_2 - d_1} \cdot C_1 \cdot C_2 \tag{3.23}$$

where

DS	is the dynamic stability of the asphalt mixture (time/mm).
t_1 and t_2	are the test times, usually 45 min and 60 min, respectively.
d_1 and d_2	are the deformations of the sample surface corresponding to test times t_1 and $t_{2,\ respectively}$.
N	is the round–trip walking speed of the test wheel, usually 42 times/min.
C_1 and C_2	are the correction coefficients of the testing machine or specimen.

According to previous studies [7], 60% of the high-temperature rutting resistance of an asphalt mixture depends on the interlocking effect of mineral aggregate particles, and 40% of the resistance depends on the adhesion of the asphalt binder. The influence of the mineral properties and graded composition on the high-temperature performance of asphalt mixtures is very important. The crushed stone aggregate has a rough surface and is angular, and the particles close to the cubic gravel aggregate have a dense skeleton structure. After compaction, the aggregate particles can form a tight locking function, which increases the internal friction angle and high-temperature stability of the asphalt mixture. The higher the high-temperature viscosity of the asphalt is, the better the adhesion with the aggregate, and the stronger the resistance of the asphalt mixture to high-temperature deformation. Suitable modifiers can be used to improve the high-temperature viscosity of asphalt, reduce the temperature sensitivity and improve the adhesiveness of the asphalt mixture, thus improving its high-temperature stability.

(2) Low-temperature crack resistance of asphalt mixtures

When the temperature decreases, the asphalt pavement surface course will experience volume contraction, and under the constraints of the base structure and surrounding materials, the asphalt mixture cannot be free to contract, which will produce temperature stress in the structural course. Because the asphalt mixture has a certain stress relaxation ability, when the cooling rate is slow, the temperature stress gradually decreases with increasing time, which does not cause great harm to the asphalt pavement. However, when the temperature drops sharply, the resulting temperature stress is too late to relax. When the temperature stress exceeds the allowable stress value of the asphalt mixture, the asphalt mixture will crack, resulting in cracks in the asphalt pavement. Therefore, the asphalt mixture must have a certain low-temperature crack resistance.

At present, the tests used to evaluate the low-temperature crack resistance of asphalt mixtures include failure tests under equal strain loading (such as indirect tensile tests and direct tensile tests), low-temperature shrinkage tests, low-temperature creep bending tests, temperature stress tests of restricted samples and stress relaxation tests.

(3) Fatigue characteristics of the asphalt mixture

The fatigue failure of an asphalt mixture means that under the action of repeated stress, it will be destroyed when it is lower than the ultimate stress under one static load. In the process of use, asphalt pavement is subjected to the repeated action of vehicle loading or temperature stress caused by the alternating change in ambient temperature, and it is in a state of repeated changes in stress and strain for a long period of time. With increasing load, the internal defects and microcracks of the materials continue to expand, and the strength of the pavement structure gradually decreases until fatigue damage occurs and cracks appear on the pavement.

At present, the common methods used to test the fatigue of asphalt mixtures worldwide include splitting fatigue tests, trapezoidal cantilever beam bending methods and rectangular beam four-point bending methods. The American SHRP A-003A research project evaluated and analyzed the sensitivity reliability and rationality of these three test methods, comprehensively considered the requirements of specimen making and test operation, and finally determined the four-point bending fatigue test of rectangular beams as the standard test for the fatigue performance of asphalt mixtures.

(4) Durability of the asphalt mixture

Durability refers to the ability of an asphalt mixture to resist environmental factors and the repeated action of traffic loads during the service period, which includes the comprehensive properties of the asphalt mixture, such as aging resistance, water stability and fatigue resistance.

1. Aging resistant properties

During the service period of an asphalt mixture, it is exposed to oxygen, water and ultraviolet rays in the air, which leads to complex physical and chemical changes in the mixture, which makes the asphalt mixture brittle and easy to crack, resulting in various cracks in the asphalt pavement.

The aging of an asphalt mixture depends on the degree of aging of the asphalt, which is related to environmental factors and the compaction void rate [8]. The greater the void ratio of the asphalt mixture is, the stronger the effect of environmental factors on the asphalt, and the higher the degree of aging. In asphalt pavement engineering, to slow the aging speed and degree of asphalt, anti-aging asphalt should be selected. During the construction of an asphalt mixture, the mixing and heating temperatures should be controlled, and the compaction density of the asphalt pavement should be ensured to reduce the aging rate of the asphalt during the construction and service periods. Considering only the durability, a fine-grained dense graded asphalt mixture can be selected, and the amount of asphalt should be increased to reduce the void ratio of the asphalt mixture to prevent water infiltration and reduce the aging effect of sunlight on asphalt materials.

2. Water stability

Owing to the effect of water, asphalt peels off from the surface of aggregate particles, which reduces the adhesive strength of the asphalt mixture. Loose aggregate particles are removed by wheels, and potholes of different sizes are formed on the road surface, resulting in water damage to the asphalt pavement. When the compaction void ratio of the asphalt mixture is high and the drainage system of the asphalt pavement is not perfect, the water trapped in the pavement structure will soak the asphalt mixture for a long time, and the dynamic water pressure caused by driving will have a peeling effect on the asphalt, which will aggravate the degree of water damage to the asphalt pavement.

Tests such as the boiling method and water immersion method can preliminarily evaluate the adhesion between asphalt and aggregate and should also be combined with the results of the water stability of the asphalt mixture test to provide a comprehensive evaluation. The tests used to evaluate the water stability of asphalt mixtures include immersion tests and freeze–thaw splitting strength tests. Immersion tests include immersion Marshall tests, immersion rutting tests, immersion splitting strength tests and immersion compressive strength tests. Under immersion, the overall mechanical strength of the asphalt mixture decreases due to the decrease in the adhesion between the asphalt and aggregate. The water stability of the asphalt mixture is evaluated by the Marshall stability ratio, rutting depth ratio, splitting strength ratio, and compressive strength ratio before and after immersion. The test conditions of the freeze–thaw splitting strength test are stricter than those of the general immersion test, and the test results are more consistent with the actual situation.

(5) Skid resistance of asphalt pavement

The skid resistance of asphalt pavement is crucial for driving safety, and it must be ensured through the reasonable selection of asphalt mixture materials and correct design and construction. The skid resistance of asphalt pavement is closely related to the surface structural depth, particle shape and size, and polishing resistance of the mineral materials. The depth of the mineral surface structure depends on the mineral composition, chemical composition, and degree of weathering. The shape and size of the particles are not only influenced by the mineral composition but also related to the processing method of the mineral materials. In addition to all the above factors, the polishing resistance is also affected by the hardness of the mineral components. Therefore, the coarse aggregate used on the surface of asphalt pavement should be gravel or crushed gravel aggregate with a rough surface, hardness, wear resistance, and large polishing value. Usually, hard and wear-resistant mineral materials are mostly acid stones, which have poor adhesion to asphalt. To ensure the water stability of the asphalt mixture, effective anti-peeling measures should be taken.

3.4 Cement and Cement Concrete

3.4.1 Cement

Cement is a water-hardened inorganic cementitious material. Cement is mixed with water to form a plastic paste, which is then transformed into a hard, stone-like solid through a series of physicochemical reactions. In terms of hardening conditions, cement not only hardens in air but also hardens better in water, maintaining and continuing to develop its strength. Therefore, cement materials can be used for both above- and below-ground work.

There are many different types of cement, which are classified according to the hydraulic minerals present in the cement: Portland cement, aluminate cement, sulfate cement, and ferro-aluminate cement, among others. Cements are classified according to their use and properties: common cements and special cements. Common Portland cements are those commonly used in general civil engineering and construction projects, such as Portland cement, ordinary Portland cement, and slag Portland cement. Special cements are cements with special uses and properties and are named after the main mineral names, properties or uses of the cement, such as sulfate aluminate cement and low heat slag Portland cement.

I. **Material composition of Portland cements**

Portland cement is a water-hardened cement composed of Portland cement clinker, a suitable amount of gypsum, and a specified amount of mixed materials. It can be divided into six varieties, namely Portland cement, ordinary Portland cement, Portland-Slag cement, Portland-Pozzolana cement, Portland-Fly Ash cement, and Portland-Composite cement, based on the variety and quantity of blended materials added to the cement. The varieties, designations, and component compositions of the above common Portland cement are summarized in Table 3.3.

II. **Main minerals in the cement clinker**

Table 3.3 Types, designations, and components of common Portland cement

Type	Code	Component (% by mass)				
		Clinker and gypsum	Granulated blast furnace slag	Pozzolana materials	Fly ash	Limestone
Portland Cement	P·I	100	–	–	–	–
	P·II	≥ 95	≤ 5	–	–	–
		≥ 95	–	–	–	≤ 5
Ordinary Portland Cement	P·O	≥ 80 and < 95	> 5 and ≤ 20			
Portland-Slag Cement	P·S·A	≥ 50 and < 80	> 20 and ≤ 50	–	–	–
	P·S·B	≥ 30 and < 50	> 50 and ≤ 70	–	–	–
Portland-Pozzolana Cement	P·P	≥ 60 and < 80	–	> 20 and ≤ 40	–	–
Portland-Fly Ash Cement	P·F	≥ 60 and < 80	–	–	> 20 and ≤ 40	–
Portland-Composite Cement	P·C	≥ 60 and < 80	> 20 and ≤ 50			

The Portland cement clinker is a water-hardened cementitious substance consisting of silicate as the main mineral and is obtained by partially melting and cooling the prepared raw material. It contains no less than 66% calcium silicate minerals and has a calcium oxide-to-silicon oxide quality ratio of 2.0. The four main minerals in the cement clinker are as follows: tricalcium silicate ($3CaO \cdot SiO_2$, C3S), dicalcium silicate ($2CaO \cdot SiO_2$, C2S), tricalcium aluminate ($3CaO \cdot Al_2O_3$, C3A), and tetra calcium ferro aluminate ($4CaO \cdot Al_2O_3 \cdot Fe_2O_3$, C4AF).

In Portland cement clinkers, the above four mineral compositions typically account for more than 95% by mass, with C3S and C2S comprising approximately 75% and C3A and C4AF accounting for approximately 22%. In addition, there are small amounts of free calcium oxide and alkali-containing minerals, such as magnesite crystals (crystalline magnesium oxide). The cement clinker produced in accordance with the regulations contains no less than 66% silicate minerals (mass fraction) and a mass–mechanical ratio of calcium oxide to silicon oxide of not less than 2.0.

III. **Physical properties of common Portland cements**

(a) Setting time

The setting time is the period required for a standard-consistency cement paste to lose its plasticity or reach a hardened state from the time it is mixed with water. The cement setting time is expressed as the time required for a standard test pin to sink into the standard consistency of cement paste to a certain depth, divided into initial and final setting times. The initial setting time is the time elapsed from the moment the cement is fully mixed with water to the initial setting state; the final setting time is the time elapsed from the moment the cement is fully mixed with water to the final setting state. The setting state of the cement is determined via a Vicatometer. The initial setting state is defined as the state of consistency when the test pin is free to sink into the test piece of standard cement paste up to 4 mm $\pm$ 1 mm from the base plate; the final setting state is defined as the state of consistency when the test pin sinks 0.5 mm and its ring attachment does not leave a trace on the surface of the test piece.

The national standard stipulates that the initial setting time of Portland cement is not less than 45 min and that the final setting time is not greater than 390 min; the initial setting times of ordinary cement, slag cement, volcanic ash cement, fly ash cement and composite cement are not less than 45 min, and the final setting time is not greater than 600 min.

The setting time of cement has important implications for the construction of cement concrete. When the initial setting time is too short, the concrete mixing and transportation, pouring and other construction processes to carry out the normal process are affected. Once the construction is completed, the concrete is required to harden as soon as possible and have a certain strength to speed up mold turnover and shorten the maintenance time. Therefore, the initial setting time of the cement should not be too short, and the final setting time should not be too long.

The setting time of cement is influenced by the type of cement and the moisture content of the cement paste. The setting time of blended cements is generally longer. If the water content of the cement paste is higher than that specified in the standard tests, the setting time is extended accordingly. In practice, the setting time of cement concrete and mortar is often much longer than that of a cement paste of standard consistency. The setting time is also influenced by the ambient temperature, which increases the rate of hydration and shortens the setting time.

(b) Soundness

Soundness is a performance indicator used to characterize whether a cement paste undergoes uneven volume changes after it has hardened. The soundness of the cement is tested via either the Le Chatelier method or the test cake method. The Le Chatelier method involves determining the standard consistency of cement paste in a ray's folder after boiling, and the relative displacement of the test needle is used to characterize the degree of volume expansion. The test cake method involves observing the standard consistency of the cement paste test cake after boiling to characterize its volume stability.

The poor volumetric stability of cement is caused by the action of certain harmful components in dry cements, which continue to react chemically with water or surrounding media after the cement paste has hardened and the volume of its products increases, causing uneven volume changes within the cement stone and generating stresses in the structure. When the stress exceeds the strength of a material, problems such as cracking and crumbling of the structure can occur. Such stresses, even if they do not exceed the strength of the cement stone, can damage the internal structure of the cement stone due to internal stress concentrations, forming defects and posing serious potential problems. The main causes of the poor volumetric stability of cement are the high free calcium oxide or magnesium oxide content in the cement clinker or the high sulphur trioxide content in the cement due to excessive gypsum admixture.

In addition, to achieve comparable results in the determination of setting time and stability, a cement paste of normal consistency should be used in both setting time and stability tests.

(c) Fineness

Fineness is an indicator of the coarseness of the cement particles or the dispersion of the cement, which affects the rate of hydration and hardening of the cement, the water requirement of the cement, the compatibility, the rate of heat release, and the strength. As the reaction between the cement and water starts on the surface of the cement particles, the finer the particles are, the larger the surface area for the cement to react with water, the more fully hydrated, and the faster the hydration rate. Therefore, for the same mineral composition of cement, the greater the fineness, the faster the setting speed and the higher the early strength. Practice has shown that increased fineness can make cement concrete stronger and work better. However, the increased fineness of cement results in greater shrinkage when it hardens in air and increases energy consumption and grinding costs. Therefore, the fineness of cement should be controlled within reasonable limits.

The fineness of the cement was tested via sieve analysis and specific surface area methods. The fineness of Portland cement and ordinary cement is expressed in terms of specific surface area, which is not less than 300 m^2/kg. The fineness of the slag cement, volcanic ash cement, fly ash cement and composite cement was tested via sieve analysis, with a sieve allowance of no more than 10% for 80 μm square sieves or no more than 30% for 45 μm square sieves.

IV. **Strength of common Portland cements**

Cement strength is an important indicator for evaluating the quality of cement and determining its strength level, as well as an important calculation parameter for the design of cement concrete and mortar ratios. In addition to the mineral composition and fineness of the cement clinker, the strength of the cement is also related to the W/C ratio, test piece production method, maintenance conditions, and time.

(a) Cement strength

To test the strength of cement, the cement can be made into cement paste, mortar, or cement concrete samples. The mortar method is currently used internationally as the standard test method for assessing cement strength.

When the cement is mixed with volcanic ash, the fluidity of cement mortar may vary considerably at an ash-to-sand ratio of 1:3 and a water-to-cement ratio of 0.5. Therefore, for volcanic ash cements, fly ash cements, composite cements, and ordinary Portland cements mixed with volcanic ash, the water requirement is determined by a water-to-cement ratio of 0.5 and a cement-sand flow of no less than 180 mm when the cement-sand strength test is carried out. When the flow of the cement sand is less than 180 mm, the water–cement ratio should be adjusted to a flow of not less than 180 mm by decreasing the flow in integral multiples of 0.01 mm, and the cement–mortar flow test should be carried out.

(b) Cement strength grade

Strength Grade of Cement is classified according to the compressive and flexural strengths measured at the specified age. The strength grades of slag cement, volcanic ash cement, fly ash cement and composite cement are divided into six grades: 32.5, 32.5R, 42.5, 42.5R, 52.5 and 52.5R.

The strength of different varieties and grades of common Portland cement should not be less than the values specified in Table 3.4 at different ages. Cements are divided into two categories on the basis of their 3D strength: normal strength and early strength (or R-type). Early-strength cements have a 3 d compressive strength of approximately 50% of the 28 d compressive strength and are more than 10% stronger than ordinary cements of the same strength class.

Table 3.4 Strength requirements of common Portland cement at different ages

Type	Grade	Compressive strength (MPa)		Flexural strength (MPa)	
		3d	28d	3d	28d
Portland Cement	42.5	17.0	42.5	3.5	6.5
	42.5R	22.0		4.0	
	52.5	23.0	52.5	4.0	7.0
	52.5R	27.0		5.0	
	62.5	28.0	62.5	5.0	8.0
	62.5R	32.0		5.5	
Ordinary Portland Cement	42.5	17.0	42.5	3.5	6.5
	42.5R	22.0		4.0	
	52.5	23.0	52.5	4.0	7.0
	52.5R	27.0		5.0	
Portland-Slag Cement Portland-Pozzolana Cement Portland-Fly Ash Cement Portland-Composite Cement	32.5	10.0	32.5	2.5	5.5
	32.5R	15.0		3.5	
	42.5	15.0	42.5	3.5	6.5
	42.5R	19.0		4.0	
	52.5	21.0	52.5	4.0	7.0
	52.5R	23.0		4.5	

3.4.2 *Cement Concrete*

Cement concrete is a mixture of cement, water and coarse and fine aggregates (also known as stone and sand) in the right proportions and is mixed with the right dosage of admixtures, dopers or other modified materials if necessary. Among them, cement plays the role of cementing and filling, and aggregates play the role of skeleton and density. The hydration reaction between cement and water produces a cementitious hydride, which firmly binds the aggregate particles into a whole, and after a set setting and hardening time, it forms an engineering composite, often referred to as concrete.

Cement concrete is widely used, is one of the most widely used materials in various buildings and structures, and has the following characteristics.

(a) A simple process with strong applicability can be poured into different shapes of the whole structure or prefabricated components according to the requirements of the engineering structure.
(b) Concrete and steel have a good grip, and steel has essentially the same coefficient of linear expansion and can be made of steel mixed
(c) High compressive strength and good durability.
(d) When the composition of the material species and ratio are changed, concrete with different physical and mechanical properties can be made to meet the requirements of different projects.

Cement concrete pavement structures are characterized by high strength, stiffness, and long service life and can withstand heavy traffic. Its main disadvantages are its high self-weight, low tensile strength, low toughness, and poor impact resistance, but this can be improved by the formulation of reinforcements and the addition of fibrous materials.

Workability of the Concrete Mixtures

The concrete mixture is a mixture of cement, water, and coarse and fine aggregates obtained after mixing. Fresh concrete is a type of concrete that has not yet set and hardened for use in the construction process, and is in a transitional state during the concrete production process. This intermediate state, from the mixing of the concrete material with water to the setting of the concrete, can be referred to as fresh concrete. The nature of fresh concrete affects both the quality of the construction of the pour and the development of the concrete properties.

The workability of concrete mixtures, also known as workability, refers to the ease with which concrete mixtures can work (mixing, transporting, pouring, vibrating, and surface preparation) and obtain uniform quality and densely formed properties. These properties largely govern the technical performance of hardened concrete, so it is highly important to study the workability of concrete and the factors influencing their workability.

The construction and ease of construction of concrete mixtures are comprehensive technical properties, including fluidity, compaction, adhesion, and water retention.

Fig. 3.6 Relationships between the compressive strength and compactness of cement concrete

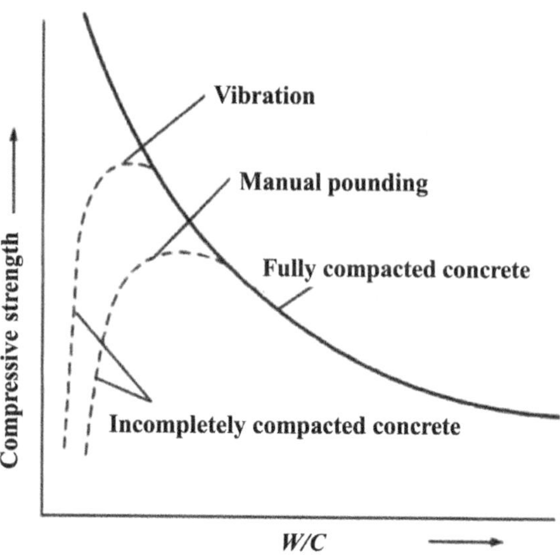

Mobility refers to the concrete mix under self-weight or mechanical pounding, which can produce flow and uniformly and densely improve the formwork performance. Compaction refers to the nature of concrete mixtures that are easily pounded and compacted to exclude all air being carried. Figure 3.6 shows the relationship between the compressive strength of cement concrete and its degree of compactness. Under the same material composition conditions, the strength of the concrete is greater after full compaction and dense forming. Cohesion refers to the fact that concrete mixtures have a certain degree of cohesion between their constituent materials during construction and do not produce delamination and segregation.

The segregation makes the distribution of the concrete composition less homogeneous, such as the settling of coarse aggregates, which leads to the separation of coarse aggregates from the paste. Water retention refers to the ability of the concrete mixture to retain water during the construction process and not to produce severe water leakage. The phenomenon of water gradually precipitating out onto the surface of the concrete mixture during construction, due to the lack of water retention, is known as waterlogging. The water will form water channels inside the concrete, making the concrete less dense and less durable.

The main influencing factors of concrete mixtures and ease of use are the material composition of the concrete and the environmental factors of construction.

Strength Characteristics of Hardened Concrete

Concrete structures are subjected to various loads and must have sufficient strength. In addition, the durability of concrete, such as frost resistance and wear resistance, is closely related to its strength. Therefore, strength is the most important mechanical property of cement concrete and a key indicator for assessing the quality of the concrete. The strength of concrete is typically referred to as its compressive strength, which is the amount of stress calculated by subjecting a standard sample to a pressure load until it is destroyed. The determination of strength requires a high degree of repeatability but allows for a certain amount of error. The strength of concrete also includes tensile, flexural, and shear strengths. These strengths are more difficult to determine accurately and are often derived from compressive strength values via empirical formulas. Other strengths, including fatigue strength and impact strength, are important for the application of specific concrete structures. The strength of concrete is influenced by several factors, the primary ones being the composition ratio, including the water-to-cement ratio, the aggregate-to-cement ratio, and the degree of compaction of the concrete.

I. **Strength of concrete**

(a) Cubic compressive strength

A cube sample with a side length of 150 mm is made according to standard methods and cured under standard conditions to an age of 28 d. The compressive ultimate breaking load is determined by standard methods, and the compressive strength of the concrete is calculated according to Eq. (3.24). The cube compressive strength of concrete is commonly used for quality control in construction work.

$$f_{cu} = \frac{F}{A} \tag{3.24}$$

where

f_{cu} is the compressive strength of the concrete (MPa).
F is the ultimate breaking load in the compression tests (N).
A is the load-bearing area of the sample (mm^2).

The standard value of the cubic compressive strength $f_{cu,k}$ for concrete is the mean value of the overall distribution of the compressive strength of the cube samples with a side length of 150 mm, which were generated and cured according to standard methods measured via standard test methods, minus 1.645 times the standard deviation. The standard value of strength $f_{cu,k}$ is guaranteed to be not less than 95%, i.e., the percentage of the overall distribution of concrete strengths below $f_{cu,k}$ does not exceed 5%. The standard value of the cube compressive strength is calculated via Eq. (3.25).

$$f_{cu,k} = \overline{f} - 1.645\sigma \tag{3.25}$$

where

$\overline{f}$ is the average of the distribution of the compressive strength (MPa).
σ is the standard deviation of the distribution of the compressive strength (MPa).

(b) Axial compressive strength

The compressive strength of concrete is determined via cubic samples, but in practice, most reinforced concrete structures take the form of prisms or cylinders. To reflect the actual force conditions more realistically, the axial compressive strength of concrete is used as the design index when calculating axially compressed members in the design of reinforced concrete structures.

The axial compressive strength is determined for specimens of size 150 mm × 150 mm × 300 mm, which better reflects the actual stresses in the concrete structure than cubes in the tests. The test results reveal that the ratio of the axial compressive strength to the cubic compressive strength is 0.7–0.8 in the range of 10–55 MPa for the cubic compressive strength.

(c) Flexural strength

In road and airport projects, concrete structures are primarily subjected to flexural loads, making flexural strength a crucial indicator for the design and quality control of concrete structures. Conversely, compressive strength serves as a reference strength indicator. The standard specimen for the flexural strength of road cement concrete is a 150 mm × 150 mm × 550 mm right-angled prismatic beam, which is cured for 28 d under standard conditions and then tested by four-point loading, and the strength of the concrete is calculated according to Eq. (3.26).

$$f_{cf} = \frac{FL}{bh^2} \tag{3.26}$$

where

f_{cf} is the flexural strength of the concrete (MPa).
F represents the ultimate breaking load in the flexural tests (N).
L is the distance between the supports (mm).
b is the width of the samples (mm).
h is the length of the samples (mm).

(d) Splitting tensile strength

The tensile strength of concrete is low, usually 1/20 to 1/10 of the compressive strength, and this ratio decreases as the compressive strength of the concrete increases. In the design of ordinary reinforced concrete structures, although the concrete is not considered to withstand tensile strength, the tensile strength of the concrete also plays an important role in the resistance to cracking and is sometimes also used to indirectly measure the bond strength of the concrete and reinforcement or for predicting concrete members due to dry shrinkage or temperature shrinkage caused by the constraints of the crack.

Tests have shown that, owing to local damage near the direct tensile strength fixture and eccentric forces, the sample is susceptible to bending, and the test results fluctuate considerably. At present, the splitting tensile test is often used to indirectly determine the tensile strength of concrete, which is referred to as the splitting tensile strength. The splitting tensile strength is calculated from Eq. (3.27).

$$f_{ts} = \frac{2F}{\pi A} = 0.637\frac{F}{A} \tag{3.27}$$

where

f_{ts} is the splitting tensile strength of the concrete (MPa).
F represents the breaking load of the concrete (N).
A is the area of the specimen splitting surface (mm^2).

II. Factors affecting the strength of concrete

When concrete is damaged by force, the rupture surface may appear at three locations. The first is damage at the interface between the aggregate and cementite, which is the most common form of damage to concrete. The second is damage to cementite, which is uncommon and occurs mainly in low-strength concrete. The third is the rupture of the aggregate itself, which occurs mostly in high-strength concrete. This analysis revealed that the strength of ordinary cement concrete mainly depends on the strength of the cement and its interfacial bonding strength with the aggregate, whereas the strength of the cement and its interfacial bonding strength with the aggregate are closely related to the composition of the concrete materials and are affected by the quality of construction, maintenance conditions and aging.

(a) Influence of the concrete composition materials

In particular, the quality and ratio of cement, water, sand, aggregate, admixtures, and other materials are the main internal factors affecting the strength of the concrete and play important roles in development.

The strength of concrete mainly depends on its internal cementitious role in the quality of cement paste, the quality of the cement paste, the strength of the cement and the water–cement ratio. When the test conditions are the same, at the same W/C ratio, the greater the strength of the cement, the greater the strength of the prepared concrete.

The strength of concrete depends on the water-to-cement ratio when the type of cement is certain. Theoretically, the W/C ratio required for full hydration of the cement is approximately 0.23, but the concrete mixture with this W/C ratio will be too dry and hard to compact under certain pounding conditions. Additionally, in partially compacted concrete, there are more pores, which reduce the strength. To achieve the necessary fluidity, when the concrete mixture is actually mixed, more water is typically added, resulting in a larger W/C ratio. When the amount of water used is too large, even if the concrete is fully compacted, after the concrete has hardened, some water remains in the concrete to form blisters or pores after evaporation, thus greatly reducing the effective cross-section of the concrete to resist the load and potentially generating stress concentrations around the pores. Therefore, for the same cement strength, the strength of the concrete decreases with an increasing water-to-cement (W/C) ratio.

Notably, the W/C ratio rule is valid only within a certain range. At very low W/C ratios, the relationship between the strength and W/C ratio may be reversed because the concrete is not sufficiently dense. In addition, the type of cement, the degree of hydration of the cement, its physical and chemical properties, the temperature during the hydration reaction, the gas content of the concrete, and variations in the effective water-to-cement (W/C) ratio all influence the W/C ratio.

When the concrete is stressed, tensile and shear stresses are generated at the interface between the coarse aggregate and the mortar. If the interface bonding strength is guaranteed, the stresses on the coarse aggregate particles are greater than those on the mortar. If the aggregate strength is insufficient, the concrete may be damaged by the destruction of the coarse aggregate. In general, the strength of the aggregate is greater than that of the cement (with the exception of light aggregates), so it does not directly affect the strength of the concrete. However, if the strength of the aggregate decreases due to weathering, the strength of the soil mixture prepared with it also decreases.

The particle shape, surface characteristics, and surface cleanliness of coarse aggregates affect the strength of the interface between them and the mortar and are important factors in determining the strength of concrete. The use of aggregates with a high content of needle-like particles not only has a negative impact on construction but also increases the void ratio of the concrete, thus reducing its strength. Crushed stone has a rough surface and is less fluid in concrete mixtures with the same amount of cement and water, but binds well to the cement mortar and therefore has a greater strength. Pebbles, on the other hand, are mostly smooth spherical particles, so concrete mixes made with pebbles have better fluidity but poorer bonding strength. When the liquidity is the same, the use of pebbles can appropriately reduce the unit water consumption of the concrete mixture; under these conditions, the strength of the hardened pebble concrete is not necessarily lower than that of concrete with crushed stone. The impurities covering the surface of the aggregate, such as silt, clay, and humus, reduce the interfacial bonding strength and affect the strength of the cement. The aggregate shape, surface configuration, and cleanliness have greater effects on the flexural strength of concrete than on the compressive strength of concrete.

The maximum size of the coarse aggregate affects both the compressive strength and the bending strength of the concrete, but the degree of effect varies. Under certain proportional conditions, the maximum particle size of the aggregate is too large, which reduces the total area of contact with the cement paste, weakens the interface strength, and compromises the concrete's strength due to inadequate compaction. This effect is more pronounced when the water-to-cement ratio is small and has a greater impact on the flexural strength of the concrete than on its compressive strength.

The advantage of continuous grading is that it produces denser concrete with excellent workability and is less prone to segregation. Intermittent grading requires less cement to prepare the same strength of concrete, but is prone to segregation.

When mixing concrete, the grain gradation and coarseness of the sand should be considered simultaneously. When the sand contains more coarse sand and the appropriate medium sand and less fine sand to fill its voids, the sand void ratio and total surface area are smaller, which is more ideal for grading, not only for cement paste with less machinery but also for improving the compactness and strength of the concrete.

(b) Curing conditions

To obtain high-quality concrete, it must be cured in a suitable environment after forming, with the aim of ensuring that the hydration process of the cement proceeds properly. For a given concrete, the rate and degree of hydration of the cement, as well as the structural characteristics of the hydrate, are dependent on the temperature and humidity of the curing environment.

The development of strength in concrete when cured in water at different temperatures has important relationships with the curing temperature of the concrete. When the curing temperature is high, the initial hydration rate of the cement increases, resulting in a higher early strength of the concrete. However, the higher the early curing temperature is, the lower the rate of strength enhancement in the later stages of concrete, which is due to the rapid early hydration reaction, resulting in an uneven distribution of cement hydrates. Areas of sparse hydration become weak points in the cement, whereas in areas of dense hydration, the hydration wraps around the cement particles, preventing further hydration of the cement and thus reducing the degree of hydration. At relatively low curing temperatures, the hydration of the cement is slower, allowing the hydrides sufficient time to diffuse and become evenly distributed throughout the cement, which leads to an increase in the concrete's strength at a later stage. However, if the curing temperature of the concrete is too low or drops below the freezing point, the hydration of the cement stops, resulting in the concrete no longer developing in strength and possibly losing strength due to the freezing effect.

Water is a necessary component of the hydration reaction of cement. If the humidity is insufficient, the hydration reaction of the cement will not proceed properly or even stop. This significantly reduces the strength of the concrete and loosens the structure of the cement, resulting in dry cracks and compromising the durability of the concrete. Concrete that is cured in air has lower strength values at all ages. Therefore, for the concrete to harden properly, conditions should be created to maintain a constant moist environment during the concrete curing period, thus producing more hydration products and increasing the compactness of the concrete. Particular attention should be given to curing in the summer due to the high temperatures and rapid evaporation of water.

The pattern of increasing concrete strength with age. Under standard curing conditions, there is a good correlation between the concrete strength and age, typically represented by a straight line on a logarithmic scale. During the construction of concrete, the early strength of the concrete can be used as a basis for extrapolating the later strengths, so that when the early strength of the concrete is insufficient, timely measures can be taken to ensure the quality of the concrete and avoid losses.

In addition to the above factors, the strength of concrete is also influenced by additives, maintenance methods, and construction methods, among others.

Durability of Hardened Concrete

Durability refers to the ability of concrete to resist the action of the surrounding environmental media to maintain its quality in use. Most concrete work is permanent, so concrete is required to have good durability under the environmental conditions in which it is used.

I. Concrete impermeability

The resistance of concrete to liquid or gas penetration is referred to as its impermeability. In a concrete structure project, the main environmental factors affecting the quality of concrete used include the following: freshwater dissolution and sulfate chemical erosion, which can cause a reduction in the strength of the cement stone. The role of carbon dioxide, chlorine, and oxygen, among others, results in the corrosion of reinforcing steel in concrete. The cracking of the concrete and other damage causes the alkali aggregate reaction. As various erosive media enter the concrete through infiltration, infiltration performance is an important factor affecting the durability of the concrete.

The permeability rating indicates the permeability of concrete. The standard samples are cured for 28 d and tested according to the prescribed method. The maximum water pressure that the concrete can withstand can be divided into six grades: S2, S4, S6, S8, S10, and S12, which indicate that the concrete can resist 0.2, 0.4, 0.6, 0.8, and 1.0, and 1.2 MPa of water pressure, without water penetration.

II. Frost resistance of the concrete

The frost resistance of concrete refers to the ability of concrete to resist the effects of freeze–thaw cycles. In cold regions, concrete under wet conditions undergoes a freeze–thaw cycle, which reduces the strength, density, and modulus of elasticity of the concrete.

The frost resistance of concrete is generally expressed in terms of the frost resistance mark. The frost rating is determined by the maximum number of freeze–thaw cycles that a standard sample of 28 d of age can withstand after saturation with water at temperatures ranging from − 20 to −15 °C to 15–20 °C to achieve a reduction in compressive strength of no more than 25% and a loss of mass of no more than 5%. The concrete frost resistance grades are D10, D15, D25, D50, D100, D150, D200, D250, and D300, which are nine grades, indicating that the concrete can withstand 10, 15, 25, 50, 100, 150, 200, 250, and 300 repeated freeze–thaw cycles. The frost resistance of concrete can also be expressed in terms of the maximum number of cycles that it can withstand if the relative modulus of elasticity value is not less than 60% and if the mass loss rate does not exceed 5%.

III. Chemical erosion resistance of concrete

The chemical attack of concrete by environmental media includes freshwater attack, seawater attack, acid attack, alkali attack, etc. The mechanism of attack is the same as that of chemical attack on cement. In addition to sulfate erosion, seawater erosion also involves repeated wet and dry cycles, the crystallization and aggregation of salts in the concrete, wave impact abrasion, and the rusting of reinforcement by chloride ions in seawater, which can also cause damage to the concrete through erosion.

In summary, the permeability, frost resistance, and chemical resistance of concrete are interrelated and are related to the degree of compactness of the concrete, i.e., the total number of pores and the structural characteristics of the pores. If the pores within the concrete form interconnected channels for water penetration, the concrete will exhibit poor permeability, and its corresponding resistance to frost and chemical attack will be reduced. Therefore, effective measures should be taken to improve the void structure of concrete and reduce the capillary channels within it to decrease the permeability of concrete, thereby enhancing its frost and chemical attack resistance. The commonly used methods include the use of water-reducing agents to reduce the water–cement ratio to increase the density of the concrete, strengthening maintenance to eliminate construction defects, preventing the formation of void channels in the concrete by dry segregations, water secretion, etc. In addition, the water-reducing agents can also be mixed with air-entraining agents in the concrete to form a uniform distribution of disconnected micropores, which cushions the squeezing pressure due to water freezing and improves the frost resistance of the concrete. The use of external protection measures to isolate the erosion medium from the concrete enhances the chemical resistance of the concrete, particularly against acid corrosion.

IV. **Abrasion resistance**

Wear resistance refers to the ability of concrete to resist damage to the surface layer. The surface layer of concrete on roads is subject to abrasion by vehicles, and the concrete on bridge piers and dams is subject to scouring by high-velocity water; therefore, abrasion resistance is one of the important properties of concrete structures for roads and bridges.

To evaluate the wear resistance of the concrete, a cube sample with a side length of 150 mm was used for curing to the specified age, a constant amount of 60 °C was used for baking, a wheel grinding head was used for the concrete grinding machine, and a load of 200 N grinding for 50 revolutions, according to Eq. (3.28), was used to calculate the amount of wear of the sample. The greater the amount of wear is, the worse the wear resistance of the concrete.

$$G = \frac{m_0 - m_1}{0.0125} \tag{3.28}$$

where

G is the wear per unit area of the test sample (kg/m^2).

m_0 is the mass of the sample before wear (kg).

m_1 is the mass of the sample after wear (kg).

0.0125   is the wear area of the sample (m^2).

The wear resistance of concrete is closely related to its strength level; however, the type of cement, aggregate hardness, and fine aggregate composition have a greater impact on the wear resistance of road concrete. According to foreign sources, when the quartz content in the sand exceeds 1/3, the abrasion resistance of the concrete pavement is significantly enhanced. For the general anti-wear requirements of the concrete, its strength grade should not be less than C20; for the anti-wear requirements of higher concrete, its strength grade should not be less than C30.

Discussion, Questions, and Exercises

(1) What are the advantages and disadvantages of different road materials? Explain what factors should be considered when selecting materials for actual engineering projects.

(2) Describe the three main layers of the road structure (cushion course, base course, and surface course) and their respective functions. Explain why each layer requires different material properties and discuss how these properties affect the overall performance of the road.

(3) How do the density, void ratio, and particle shape of aggregates affect the performance of road materials? A practical case is provided to illustrate how these physical properties affect the durability and load-bearing capacity of a road.

(4) The design of asphalt mixtures needs to consider various factors, including high-temperature stability, low-temperature crack resistance, water stability, and skid resistance. A specific traffic environment (such as a high-temperature and rainy area) is chosen, and an asphalt mixture is designed, which explains how to optimize the performance of the mixture to adapt to this environment by adjusting the proportions of materials and selecting additives.

References

1. ASPHALT INSTITUTE. Performance Graded Asphalt Binder Specification and Testing [M]. Asphalt Institute, 1994

2. HO C-H, ROMERO P. Asphalt Mixture Beams Used in Bending Beam Rheometer for Quality Control: Utah's Experience [J]. Transportation research record, 2012, 2268(1): 92–7

3. MORIYOSHI A, SHEN J, EZAWA K, et al. Comparison of Various Testing Methods for Low-temperature Properties of Asphalts [J]. Journal of the Japan petroleum Institute, 2005, 48(6): 336–43

4. ANDERSON D A, DONGRE R. The SHRP Direct Tension Specification Test-its Development and Use [J]. ASTM Special Technical Publication, 1995, 1241: 51–66

5. MASAD E A, HUANG C-W, DANGELO J, et al. Characterization of Asphalt Binder Resistance to Permanent Deformation Based on Nonlinear Viscoelastic Analysis of Multiple Stress Creep Recovery (MSCR) Test [J]. Journal of the Association of Asphalt Paving Technologists, 2009, 78

6. SUKHIJA M, SABOO N. A Comprehensive Review of Warm Mix Asphalt Mixtures-laboratory to Field [J]. Construction and Building Materials, 2021, 274: 121781

7. GOLALIPOUR A, JAMSHIDI E, NIAZI Y, et al. Effect of Aggregate Gradation on Rutting of Asphalt Pavements [J]. Procedia-Social and Behavioral Sciences, 2012, 53: 440–9

8. YANG B, LI H, XIE N, et al. Surface Characteristics of Ageing Asphalt Binder Coupling Thermal Oxidation and Ultraviolet Radiation [J]. Transportation Research Record, 2022: 03611981221088583

Chapter 4
Pavement Structure Design

4.1 Asphalt Pavement Design

4.1.1 Pavement Structural Layer Division and Functional Requirements

The design of asphalt pavement includes three parts: combination design, thickness design and material design [1, 2]. The pavement structure design mainly includes two parts: structure combination and thickness design. The main task of structural design is to determine the hierarchy combination, structural layer thickness, material strength and modulus according to the load and material strength and stiffness characteristics so that the stress and displacement components in the asphalt pavement structure are properly controlled within the allowable range [3–5].

The asphalt pavement structure layer can be composed of a surface layer, a base layer, a subbase layer, a cushion layer and other multilayer structures [6–8].

(1) The surface layer can be a single layer, a double layer or three layers, and the double layer structure is called the surface layer, i.e., the layer below. The three-layer structure is called the surface layer, the middle layer and the lower layer. The surface layer should have flat compactness, skid resistance, crack resistance and durability; the middle and lower layers should have high temperature resistance to rutting, shear resistance, compaction, and basic impermeable performance; and the bottom layer should be resistant to fatigue.

(2) The base layer is the main load-bearing layer, which should be stable, durable and have a high bearing capacity. The base layer can be a single or double layer. Both asphalt mixtures, granular flexible bases, semirigid bases, and rigid bases, are required to have relatively high physical and mechanical properties.

(3) The subbase layer is set under the base and the surface layer (the wheel load repeated bearing layer).

© Tongji University Press Co., Ltd. 2026

H. Li et al., *Road Engineering*, https://doi.org/10.1007/978-981-95-6659-4_4

(4) The cushion layer is arranged in the bottom layer and the structure between the soil base layer, with drainage, water insulation, antifreeze and other functions. All highway levels should set the necessary structural layer according to the specific situation.

4.1.2 Chinese Asphalt Pavement Design Indices and Standards

The design index is a control index proposed from the perspective of the mechanical response, which should cover the main types of pavement structure diseases. The design control standard refers to the limit state of the pavement structure reached according to the failure process and failure mechanism of the design index. In the design of a pavement structure, if the structure combination meets the limit state of the control index, the pavement structure can be guaranteed to work normally during the design period of use without causing the limit state of damage [9, 10].

Design Indicators

The stress distribution of each structural layer of the asphalt pavement structure under the action of wheel loading is very complex. Theoretical calculations and a large number of experimental verifications show the following:

(1) Due to the rigid plate structure effect, the ultimate tensile stress generally appears in the cement concrete base or inorganic binder stable base plate bottom. The initial cracks are generated and further developed to form fracture cracks, which induce stress redistribution in the asphalt surface layer, and the upward reflection of cracks causes surface layer failure.
(2) For the pavement structure under the inorganic binder stabilization class, the initial crack usually occurs at the bottom of the lower base and then gradually extends upward to the base and asphalt surface.
(3) For flexible base asphalt pavement, when the flexible base material is mainly asphalt binder, the bottom of the asphalt binder base will bear main tensile stress. When the flexible base material is mainly granular, the granular base material does not bear tensile stress, whereas the asphalt surface layer will bear greater tensile stress. Therefore, the limit state of flexible base asphalt pavement and the whole pavement structure mainly appears at the bottom of the asphalt mixture layer, forming initial cracks, gradually expanding, and finally forming fracture cracks in the asphalt surface layer.
(4) For the asphalt mixture layer and roadbed, irrecoverable permanent deformation occurs under the vertical compressive stress and shear stress of the track load. When cement concrete or inorganic binder is used to stabilize the base, permanent deformation mainly occurs in the asphalt mixture layer. When flexible

substrates are used, permanent deformations may accumulate over the entire range of the structure.

The selection of the design index should correspond to the main mechanical response of the asphalt pavement structure layer and be used to control the occurrence of its main diseases. After long-term observations and research of engineering around the world, it has revealed that the tensile stress in the bottom layer of a pavement structure under a wheel loading structure limits tensile stress generally occurs in the bottom layer. The tensile stress of layer structure (generally) is the first principal stress at and above the tensile strength limit of the layer material. The first initial cracks under repeated wheel loading extends gradually and expands in the vertical direction, resulting in a variety of cracks on the surface of the road. Further development will lead to damage to a larger range. At the same time, for the asphalt pavement structure, even if the residual deformation generated by each driving load is small, the sum of residual deformation accumulated by repeated actions will be large enough to affect the normal driving of vehicles. Therefore, the structural design of asphalt pavement uses the bottom tensile strain of the asphalt mixture layer, the inorganic binder stabilized layer bottom tensile stress, the permanent deformation of the asphalt mixture layer and the vertical compressive strain of the roadbed top surface as important control indicators of structural design. To control the fatigue cracking of the asphalt mixture layer, the fatigue cracking of the inorganic binder stabilized layer and the permanent deformation of the asphalt layer are used.

To prevent low-temperature cracking and freeze–thaw damage to pavement structures in seasonally frozen soil areas, the low-temperature cracking index of the asphalt surface layer and the frost-resistant thickness of the pavement structure are also important performance control indices. The low-temperature cracking index (CI) refers to the number of transverse cracks in the 100 m survey unit when the asphalt pavement is inspected. The crack that runs through the whole width is counted as crack 1, the crack that does not run through and is longer than the width of one lane is counted as crack 0.5, and the crack that does not exceed the width of one lane is not included.

Design Criteria

Under the repeated action of wheels on asphalt pavement, the tensile stress at the bottom of the asphalt surface layer and rigid and semi-rigid material layer exceeds the limit, forming initial cracks and gradually extending to fracture, which results in fatigue fracture damage. Therefore, with respect to the main asphalt pavement structures of China, the "Specifications for Design of Highway Asphalt Pavement" (JTG D50) stipulates that the tensile strain at the bottom of the asphalt mixture layer and the tensile stress of the inorganic binder layer are taken as the design indices, and the fatigue cracking life of the asphalt mixture layer and inorganic binder layer is taken as the design standard. The fatigue cracking life of the asphalt mixture layer calculated on the basis of the bottom tensile strain of the asphalt mixture layer should be greater than the cumulative action time of the equivalent design axial load within the design life calculated on the basis of the bottom tensile strain of the asphalt mixture layer. The fatigue cracking life of the inorganic binder stable layer calculated on the basis of the bottom tensile stress of the inorganic binder stable layer should be greater than the cumulative action times of the equivalent design axial load within the design life calculated on the basis of the bottom tensile stress of the inorganic binder stable layer.

For the asphalt pavement structure, even if the residual deformation generated by each driving load is small, the sum of the residual deformation accumulated by many repeated actions will be large enough to affect the normal driving of vehicles. Therefore, from the perspective of controlling the permanent deformation of the asphalt pavement structure, China's Code for Design of Highway Asphalt Pavement (JTG D50) requires that the permanent deformation of the asphalt mixture calculated on the basis of the cumulative times of the equivalent design axle load within the design life should not be greater than the allowable permanent deformation. Moreover, the vertical compressive strain on the top surface of the subgrade should not be greater than the allowable vertical compressive strain calculated on the basis of the cumulative action times of the equivalent design axial load within the design life.

4.1.3 Pavement Structure Checking Method

Temperature Adjustment Coefficient and Equivalent Temperature

Air temperature is an important external factor affecting pavement performance. The modulus of an asphalt mixture is typically dependent on temperature. However, in the current structure design of asphalt pavement in China, when the pavement structure is checked, the modulus of the asphalt mixture structure layer is fixed at the standard test temperature of 20 °C. To consider the influence of temperature, China's Code for Design of Highway Asphalt Pavement (JTG D50), according to the temperature conditions, pavement structure types and structural layer thickness in the region, the temperature adjustment coefficient is used to characterize the influence of different climatic conditions on the fatigue cracking of the pavement structure layer and vertical compressive strain of the subgrade top surface, and the equivalent temperature is used to characterize the influence on the permanent deformation of the asphalt mixture layer according to local climatic conditions.

Generally, the temperature adjustment coefficient and equivalent temperature are determined in two steps. First, the temperature adjustment coefficient and equivalent temperature of the benchmark pavement structure are determined, and then the thickness and modulus of the structural layer are modified to obtain the temperature adjustment coefficient and equivalent temperature of the pavement with different structures.

When the asphalt surface layer or base layer (including the bottom base layer) of the pavement structure is composed of two or more layers of different material structure layers, the equivalent asphalt surface layer and equivalent base layer can be converted into an equivalent asphalt surface layer and an equivalent base layer according to Eqs. (4.1) and (4.2) to simplify the pavement structure into a three-layer pavement structure composed of an equivalent asphalt surface layer, an equivalent base layer and a subgrade. For a road with an asphalt binder base, the base is converted to an equivalent asphalt layer. When there are more than two layers, Eqs. (4.1) and (4.2) are reused for layer-by-layer conversion from top to bottom, which is simplified into a three-layer pavement structure composed of an equivalent asphalt surface layer with an equivalent base and subgrade.

$$h_i^* = h_{i1} + h_{i2}$$
(4.1)

$$E_i^* = \frac{E_{i1}h_{i1}^3 + E_{i2}h_{i2}^3}{(h_{i1} + h_{i2})^3} + \frac{3}{h_{i1} + h_{i2}}\left(\frac{1}{E_{i1}h_{i1}} + \frac{1}{E_{i2}h_{i2}}\right)^{-1}$$
(4.2)

where h_i^* and E_i^* are the equivalent layer thickness (mm) and modulus (MPa), respectively, with subscript i = a for the asphalt surface and i = b for the base.

The temperature adjustment coefficient of the pavement structure should be calculated according to Eqs. (4.3)–(4.17).

$$K_{Ti} = A_h A_E \check{k}_{Ti}^{\,1+B_h+B_E}$$
(4.3)

where K_{Ti} is the temperature adjustment coefficient. The subscript i = 1 corresponds to the fatigue cracking analysis of the asphalt mixture layer, i = 2 corresponds to the fatigue cracking of the inorganic bonding layer, and i = 3 corresponds to the vertical compression and strain analysis of the subgrade top surface.

$\check{k}_{Ti}$ is the temperature adjustment coefficient of the reference pavement structure;

A_h, B_h, A_E and B_E are functions related to the surface layer, base layer thickness and modulus, which are calculated according to Eqs. (4.4)–(4.17).

Fatigue cracking of the asphalt mixture layer:

$$A_E = 0.76\lambda_E^{0.09}$$
(4.4)

$$A_h = 1.14\lambda_h^{0.17}$$
(4.5)

$$B_E = 0.14\ln(\lambda_E/20)$$
(4.6)

$$B_h = 0.23\ln(\lambda_h/0.45)$$
(4.7)

Fatigue cracking of a stable layer of inorganic binder:

$$A_E = 0.10\lambda_E + 0.89 \tag{4.8}$$

$$A_h = 0.73\lambda_E + 0.89 \tag{4.9}$$

$$B_E = 0.15\ln(\lambda_E/1.14) \tag{4.10}$$

$$B_h = 0.44\ln(\lambda_h/0.45) \tag{4.11}$$

Vertical compressive strain on the top of the subgrade:

$$A_E = 0.006\lambda_E + 0.89 \tag{4.12}$$

$$A_h = 0.67\lambda_h + 0.70 \tag{4.13}$$

$$B_E = 0.12\ln(\lambda_E/20) \tag{4.14}$$

$$B_h = 0.38\ln(\lambda_h/0.45) \tag{4.15}$$

where:

λ_E is the ratio of the equivalent modulus of the surface layer and the base layer, which is calculated according to Eq. (4.16):

$$\lambda_E = \frac{E_a^*}{E_b^*} \tag{4.16}$$

λ_h is the ratio of the equivalent modulus of the surface layer and base layer and can be calculated according to Eq. (4.17):

$$\lambda_E = \frac{h_a^*}{h_b^*} \tag{4.17}$$

When analyzing the permanent deformation of the asphalt mixture layer, the equivalent temperature of the asphalt mixture layer should be calculated according to Eq. (4.18).

$$T_{pef} = T_\xi + 0.016h_a \tag{4.18}$$

where

T_{pef} Equivalent temperature of the asphalt mixture layer (°C);
h_a Asphalt mixture layer thickness (mm);
T_ξ Reference equivalent temperature.

Checking the Calculation of the Fatigue Cracking of the Asphalt Concrete Layer

On the basis of the flexible characteristics of the asphalt mixture, the fatigue cracking life of the asphalt mixture layer is calculated and controlled by the tensile strain of the asphalt mixture layer. The results show that the fatigue cracking model of the thin asphalt mixture layer is suitable for the normal strain loading mode, the fatigue cracking model of the thick asphalt mixture layer is suitable for the normal stress loading mode, and the transition relationship between the two is needed for the intermediate thickness of the asphalt mixture layer. On the basis of the fatigue data of many accelerated loading test roads at home and abroad, the current "Code for Highway Asphalt Pavement Design" (JTG D50) has established a calculation model for the fatigue cracking life of an asphalt mixture layer on the basis of the bottom tensile strain of the asphalt mixture layer. Considering the transition and transformation of different loading modes, the fatigue cracking loading mode coefficient is introduced into the model.

$$N_{f1} = 6.32 \times 10^{1.56-0.29\beta} k_a k_b k_{T1}^{-1} \left(\frac{1}{\varepsilon_a}\right)^{3.97} \left(\frac{1}{E_a}\right)^{1.58} (VFA)^{2.72} \tag{4.19}$$

where

N_{f1} Fatigue cracking life of the asphalt mixture layer.
β Target reliability index;
K_{T1} Temperature adjustment factor;
ε_a Tensile strain at the bottom of the asphalt mixture layer (10^{-6}), calculated according to elastic layering theory;
k_a Seasonally frozen soil area adjustment coefficient;
k_b The coefficient of the fatigue loading mode is calculated according to Eq. (4.20)

$$k_b = \left[\frac{1 + 0.3 E_a^{0.43}(VFA)^{-0.85} e^{0.024 h_a - 5.41}}{1 + e^{0.024 h_a - 5.41}}\right]^{3.33} \tag{4.20}$$

E_a Dynamic compression modulus (MPa) of the asphalt mixture at 20 °C;
VFA   Asphalt saturation of the asphalt mixture (%);
h_a Thickness of the asphalt mixture layer (mm).

The fatigue cracking life of the asphalt mixture layer should be greater than the cumulative action time of the equivalent design axle load in the design lane on the basis of the tensile strain at the bottom of the asphalt mixture layer. Otherwise, the road structure should be adjusted and rechecked until the requirements are met.

Checking the Calculation of the Fatigue Cracking of the Inorganic Binder Stabilized Layer

On the basis of the semi-rigid characteristics of inorganic binder stabilization materials, the bottom tensile stress of the inorganic binder stabilization layer is generally used to calculate and control the fatigue cracking life of the inorganic binder stabilization layer. On the basis of the extensive fatigue cracking test results of four commonly used mixtures, i.e., cement-stabilized gravel, cement-stabilized gravel, cement-stabilized soil and lime fly ash-stabilized gravel, the fatigue cracking calculation model of inorganic binder-stabilized granular material and stabilized soil is established in the Code for Design of Highway Asphalt Pavement (JTG D50), as shown in Eq. (4.21). Owing to the lack of sufficient field data, it is difficult to verify the fatigue cracking model of the inorganic binder stabilized layer. On the basis of investigations of many inorganic binder-stabilized base asphalt pavement structures, the typical structure of inorganic binder-stabilized base asphalt pavement under different working conditions, including highway grade, traffic load parameters and the subgrade rebound modulus, is summarized. The damage conditions of the typical pavement structures investigated were compared with the analysis results of the above fatigue cracking model, and the onsite comprehensive correction coefficient K_C was introduced to reflect the difference between the indoor performance model and the onsite fatigue cracking damage.

$$N_{f2} = k_a k_{T2}^{-1} 10^{a - b\frac{\sigma_t}{R_s} + k_c - 0.57\beta} \tag{4.21}$$

where

N_{f2} Fatigue cracking life of the inorganic binder stabilized layer;

k_a Coefficient of function integration in seasonal frozen soil area, determined according to the table;

k_{T2} Temperature adjustment coefficient;

R_S Flexural and tensile strengths of inorganic binder-stabilized materials (MPa);

a, b The regression parameters of the fatigue test are determined according to Table 4.1;

K_C The onsite comprehensive correction coefficient can be determined according to Equations (8.73):

$$k_c = c_1 e^{c_2(h_a + h_b)} + c_3 \tag{4.22}$$

c_1, c_2, c_3 According to Table 4.2;

h_a, h_b Thickness of the asphalt mixture layer and the inorganic binder stabilization layer above the calculation point;

β Target reliability index;

σ_t The tensile stress (MPa) at the bottom of the stable layer of the inorganic binder is calculated according to elastic laminar theory.

Table 4.1 Fatigue failure model parameters of the inorganic binder stability layer

Material type	a	b
The inorganic binder stabilizes the pellet	13.24	12.52
Inorganic binder stabilized soil	12.18	12.79

Table 4.2 Field comprehensive correction coefficient (KC)-related parameters

Material type	New pavement structure layer or reconstruct the existing pavement structure layer in the main road		Remodeling project overlay	
	The inorganic binder stabilizes the pellet	Inorganic binder stabilized soil	The inorganic binder stabilizes the pellet	Inorganic binder stabilized soil
C_1	14.0	35.0	18.5	21.0
C_2	-0.0076	-0.0156	-0.01	-0.0125
C_3	-1.47	-0.83	-13.2	-0.82

The fatigue cracking life of the stable layer of inorganic binder should be greater than the cumulative action times of the equivalent design load of the design lane within the design service life, which is obtained from the axial load conversion on the basis of the tensile stress at the bottom of the stable layer of inorganic binder. Otherwise, the pavement structure combination or layer thickness should be adjusted and rechecked until the requirements are met.

Checking the Permanent Deformation of the Asphalt Mixture Layer

Considering the stress distribution at different depths of asphalt pavement and the difference in the rutting resistance of different asphalt mixture layers, the permanent deformation is calculated on a layer-by-layer basis. The difference between the accumulated value of permanent deformation of each layer and the total permanent deformation of the asphalt mixture layer is considered in the comprehensive correction coefficient K_r.

When the permanent deformation amount within the designed service life of the pavement is estimated, the cumulative action times of the equivalent design axle load on the design lane within the design service life, which are obtained via shaft load conversion on the basis of the permanent deformation index of the asphalt mixture layer, are used to calculate the permanent deformation amount. However, the maintenance and repair of pavement should be considered comprehensively in structural analysis. For projects with a large traffic volume and a high overloading ratio, pavement design with a fixed number of years is sometimes needed for rutting repairs once or more than once to calculate the permanent deformation of the asphalt mixture layer design lane equivalent axle load number of the cumulative effect for traffic within the time limit for the first time to repair the equivalent axle load number of the cumulative effect.

In accordance with the regulations of the design of asphalt pavement in China, first, each asphalt mixture layer in the pavement structure is stratified: the surface layer is 10–20 mm thick; second, the thickness of each asphalt mixture layer should not be greater than 25 mm; third, the thickness of each asphalt mixture layer should not be more than 100 mm; and fourth and second layers of the asphalt mixture should be stratified. Then, according to the rut test under standard conditions, the permanent deformation of each layer of the asphalt mixture in the rut test is obtained, and the permanent deformation of each layer and the total permanent deformation of the asphalt mixture layer are calculated according to Eq. (4.23).

$$R_a = \sum_{i=1}^{n} R_{ai} \tag{4.23}$$

$$R_{ai} = 2.31 \times 10^{-8} k_{Ri} T_{pef}^{2.93} P i^{1.80} N_{e3}^{0.48} \left(\frac{h_i}{h_0} \right) R_{oi} \tag{4.24}$$

where

R_a Permanent deformation of the asphalt mixture layer (mm);

R_{ai} Permanent deformation of layer i (mm);

n   Number of layers;

T_{pef} Equivalent temperature of permanent deformation of the asphalt mixture layer (°C);

N_{e3} The cumulative action times of the equivalent design axle load on the design lane based on the permanent deformation index of the asphalt mixture layer during the design service life or the period from the opening to the first rut maintenance;

h_i The ith layer thickness (mm);

h_o Thickness of the rut test specimen (mm);

R_{oi} When the test temperature is 60 °C, the pressure is 0.7 MPa, the loading time is 2520 cycles, and the permanent deformation of the asphalt mixture in the rut test is mm.

K_{Ri} The comprehensive correction coefficient can be calculated according to Eq. (4.25)–(4.27):

$$k_{Ri} = (d_1 + d_2 \cdot z_i) \cdot 0.9731^{z_i} \tag{4.25}$$

$$d_1 = -1.35 \times 10^{-4} h_a^2 + 8.18 \times 10^{-2} h_a - 14.50 \tag{4.26}$$

$$d_2 = 8.78 \times 10^{-7} h_a^2 - 1.50 \times 10^{-3} h_a + 0.90 \tag{4.27}$$

Z_i The thickness of layer I of the asphalt mixture (mm); the first layer is 15 mm, and the other layers are the depths between the road surface and the midpoint of the layer.

h_a Asphalt mixture layer thickness (mm); when h_a is greater than 200 mm, the thickness is 200 mm;

p_i The vertical compressive stress (MPa) of the top surface of layer I of the asphalt mixture is calculated and obtained according to elastic layered system theory.

The calculated permanent deformation of the asphalt mixture layer should meet the requirements. Otherwise, the asphalt mixture design should be adjusted until it meets the requirements. The asphalt mixture that meets the requirement of allowable permanent deformation of the asphalt mixture layer should also meet the requirement of dynamic stability of the standard rut test as required by construction technical specifications, and the stability of its permanent deformation R_0 can be used as the quality requirement and construction control index of the asphalt mixture. When the standard rut test temperature is 60 °C, the pressure is 0.7 MPa, the thickness of the sample is 50 mm, and the loading number is 2520, the dynamic stability DS of the asphalt mixture can be calculated according to the permanent deformation amount according to Eq. (4.28).

$$DS = 9365R_0^{-1.48} \tag{4.28}$$

where DS is the dynamic stability of the asphalt mixture (times/mm).

Checking the Calculation of the Vertical Compressive Strain on Top of the Subgrade

The vertical compressive strain on the top surface of the subgrade is an important design index for asphalt pavement with a granular base and an asphalt binder base. Related design methods in foreign countries generally prevent excessive permanent deformation of roadbeds by controlling the vertical compressive strain on the top surface of the roadbed and use test road or field observation data to fit the relationship between the vertical compressive strain and traffic load parameters and lack sufficient measured data. Therefore, the pavement structure data of the AASHO test road and the number of axial loads were sorted and the empirical relationship between the vertical compressive strain on the top of the subgrade and the number of axial loads on 100 kN was established. After adjustment and correction, the calculation model of the allowable vertical compressive strain on the top of the subgrade was established, as shown in Eq. (4.29).

$$[\varepsilon_z] = 1.25 \times 10^{4-0.1\beta}(k_{T3}N_{e4})^{-0.21} \tag{4.29}$$

where

$[\varepsilon_z]$ is the vertical compressive strain allowed on the top surface of the subgrade (10^{-6});

β Target reliability index;

N_{e4} On the basis of the pressure and strain indices of the subgrade top surface, the cumulative action times of the equivalent design axle load on the design lane during the design service life were calculated;

K_{T3} Temperature adjustment factor.

For the selected pavement structure, the vertical compressive strain on the top surface of the subgrade calculated via elastic layered system theory should be less than the allowable compressive strain value. Otherwise, the road structure scheme is adjusted, and the calculation is rechecked until the requirements are met.

Checking the Low-Temperature Cracking Index of the Asphalt Surface

Low-temperature cracking of asphalt pavement is a common disease in seasonal frozen soil. The relationships among the asphalt properties, pavement structure, subgrade soil type and pavement low-temperature cracking conditions in several sections in Northeast China were analyzed via an empirical method. A prediction model of the pavement low-temperature cracking index was established by referring to the Canadian Haas model.

$$CI = 1.95 \times 10^{-3} S_t \lg b - 0.075(T + 0.07 h_a) \lg S_t + 0.15 \qquad (4.30)$$

where

CI Low-temperature cracking index of the asphalt surface;

T The pavement cracking design temperature (°C) is the average of the lowest temperatures over 10 consecutive years;

S_t Under the conditions of a low pavement design temperature plus a 10 °C test temperature, the creep stiffness (MPa) of the surface asphalt bending beam was loaded for 180 s in the rheological test.

h_a Asphalt binder material layer thickness (mm);

b Subgrade type parameters, sand: b = 5, silty clay: b = 3, clay: b = 2.

The low-temperature cracking index of the asphalt surface should meet the requirements of the specified low-temperature cracking index; otherwise, the selected asphalt material should be changed until it meets the requirements.

Antifreeze Thickness Check Calculation

When the subgrade in the seasonal frozen soil area is moderately wet or wet, the maximum frozen depth of the highway for many years should be calculated according to Eq. (4.31). According to the maximum freezing depth of the highway for many

Table 4.3 Thermal property coefficients of roadbed and pavement materials

Material	Thermal physical property coefficient
Clayey soil	1.05
Powder soil	1.1
Powder sand soil	1.2
Fine grained soil sand, clay sand	1.3
Fine earthy sand (sand)	1.35
Cement concrete concrete	1.4
Asphalt binder class	1.35
Graded crushed stone	1.45
Lime fly ash stabilized material or cement stabilized granular material	1.4
Lime fly ash stabilized material soil and cement-soil	1.35

Table 4.4 Subgrade humidity coefficient

Type	Dry	In the wet	Wet
Damp coefficient	1.0	0.95	0.90

years, the antifreezing thickness of the road surface is checked according to the provisions. When the thickness of the road surface structure is less than the minimum antifreezing thickness, an antifreezing layer should be added to meet the requirements of the minimum antifreezing thickness.

$$Z_{max} = abcZ_d \tag{4.31}$$

where

Z_{max} Annual maximum freezing depth of highway (mm);

Z_d The maximum freezing depth (mm) of the land for many years was determined according to the survey data.

a The thermal physical property coefficient of each layer material of subgrade and pavement within the range of ground freezing depths is determined according to Table 4.3;

b The subgrade humidity coefficient is determined according to Table 4.4;

c The subgrade section form coefficient is determined via interpolation according to Table 4.5.

Design Acceptance Bending Value of the Pavement Structure

The adoption of a drop hammer bending sinker is generally recommended for roadbed acceptance. The load of the drop hammer bending sinker is 50 kN, and the radius

Table 4.5 Subgrade section form coefficients

Fill and dig form and height (depth)	Subgrade filling height (m)					Subgrade excavation height (m)			
	Zero fill	< 2	2–4	4–6	> 6	< 2	2–4	4–6	> 6
Coefficient of sectional form	1.0	1.02	1.05	1.08	1.10	0.98	0.95	0.92	0.90

of the load plate is 150 mm. The acceptance bending value L_g of the subgrade top surface is calculated according to Eq. (4.32). The measured representative bending value L_0 of the top surface of the subgrade should meet the requirements of Eq. (4.33).

$$l_g = \frac{176pr}{E_0} \tag{4.32}$$

where

l_g Acceptance bending settlement value of the subgrade top surface (0.01 mm);
p The load applied to the bearing plate (MPa) of the drop weight bending apparatus;
r Radius of the bearing plate (mm);
E_0 Elastic modulus of the subgrade top surface under equilibrium humidity (MPa)

$$l_0 \le l_g \tag{4.33}$$

where

l_0 The representative value of the subgrade top subsidence measured in the section is 0.01 mm, and the section is evaluated from 1 to 3 km and calculated according to Eq. (4.34):

$$l_0 = (\bar{l}_o + \beta \cdot s)K_1 \tag{4.34}$$

where

$\bar{l}_o$ is the mean value of the subgrade top subsidence measured in the road section (0.01 mm);
s The standard deviation of the subgrade top subsidence measured in the road section (0.01 mm);
β Target reliability index;
K_1 The influence coefficient of the subgrade top bending and settling humidity is determined according to local experience.

The curve settling value l_a of the road table acceptance should be calculated according to Eq. (4.35) via the design road surface structure and elastic layered system theory. The parameters of the pavement structure layer are the same as those used for checking the pavement structure. The elastic modulus of the subgrade top surface should be multiplied by the modulus adjustment coefficient k_1 under the condition of balanced humidity to coordinate the difference between the theoretical bending and measured bending.

$$l_a = p\tilde{l}_a \tag{4.35}$$

$$\tilde{l}_a = f\left(\frac{h_1}{\delta}, \frac{h_2}{\delta}, \cdots \frac{h_{n-1}}{\delta}; \frac{E_2}{E_1}, \frac{E_3}{E_2}, \cdots \frac{k_1 E_0}{E_{n-1}}\right) \tag{4.36}$$

where

$\tilde{l}_a$ Theoretical bending coefficient;

k_1 The adjustment coefficient of the elastic modulus of the subgrade top surface, inorganic binder stable base asphalt pavement and cement concrete base asphalt pavement is set as 0.5. For the granular base asphalt pavement and asphalt binder base asphalt pavement, when an inorganic binder is used to stabilize the base, the value is 0.5; otherwise, it is 1.0.

The bending value of the road table should be detected during the time of delivery (completion) of the road table. The humidity and temperature correction of the bending value should be considered during detection. The representative values of center point bending and settling of drop weight bending and settling instruments should meet the requirements of Eq. (4.37).

$$l_0 \leq l_a \tag{4.37}$$

where

l_a Bending and settling value of road meter acceptance (0.01 mm);

l_0 The representative value (0.01 mm) of the measured road table bending and subsidence in the section is calculated according to Eq. (4.38), taking 1–3 km as an evaluation section:

$$l_0 = (\tilde{l}_0 + \beta \cdot s)K_1 K_3 \tag{4.38}$$

where

$\bar{l}_o$ The mean value of bending and settling of the measured road gauge in the section (0.01 mm);

K_1 The subgrade modulus value is obtained via reverse calculation according to the measured subgrade subsidence value, and then the subgrade modulus is corrected to obtain the structural modulus value. Then, the subgrade humidity correction coefficient K_1 under the test conditions is obtained or determined according to local experience.

K_3 The influence coefficient of the bending and settling temperatures of the circuit table can be determined according to Eq. (4.39):

$$K_3 = e^{\left[9 \times 10-6(\ln E_0-1)h_a+4\times10^{-3}\right](20-T)}$$ (4.39)

where

T The measured or estimated temperature (°C) at the midpoint of the asphalt binder layer during bending and settling measurement;

h_a Asphalt binder material layer thickness (mm).

4.2 Cement Concrete Pavement Design

4.2.1 Structural Combination Design of Cement-Reinforced Concrete Pavement

Cement Concrete Pavement Board

The cement concrete pavement plate should have enough strength, durability, skid resistance, wear resistance, smoothness and other good road performance; generally, joints (in addition to joints, plate edges and corner locations) should be used without the reinforcement of ordinary cement concrete pavement boards. At the traffic load level above heavy traffic, corner steel bars can be added, and edge steel bars should be configured for positions with weak foundations, no transmission bars or other structures [11–13].

When the plane size of the slab is large or irregular in shape, underground facilities are buried under the pavement structure, and it is located in the subgrade section where uneven settlement may occur, such as high embankments, soft soil foundation, filling and excavation junction section, etc., the reinforced concrete surface with dowel bars at joints should be adopted. Other pavement types, such as continuous reinforced concrete, roller compacted concrete and steel fiber reinforced concrete, can be selected according to the applicable conditions, as shown in Table 4.6.

Ordinary concrete, reinforced concrete, roller compacted concrete or steel fiber concrete surface layer boards generally use rectangular bins, with vertical and horizontal joint separation; longitudinal and transverse joints should be perpendicularly intersected, and longitudinal joints on both sides of the transverse joint should not be mutually dislocated. The longitudinal seam spacing is determined in the range of 3.0–4.5 m according to the pavement width. The transverse seam spacing of ordinary concrete surface plates is generally 4–6 m. The aspect ratio of the surface layer plate should not exceed 1.35, and the plane size should not be greater than 25 m². The transverse seam spacing of rolling concrete or steel fiber-reinforced concrete surface

Table 4.6 Selection of other concrete pavement surface types

The surface type		Suitable conditions
Continuous reinforced concrete surface		The highway
Compound road surface	Top layer of dense asphalt mixture	A highway with extremely heavy or heavy traffic load
	Continuously reinforced concrete lower layer	
	The lower layer of ordinary concrete is set as dowel rod	
RCC surface		Second-class and below two highways
Steel fiber concrete surface		Elevation restricted sections, concrete overlay
Concrete precast block surface		Unstable subsidence section of bridgehead approach road of second-class and below highways, parking lot in service area

plates is generally 6–10 m, and that of reinforced concrete surface plates is generally 6–15 m. The aspect ratio of the surface plate should not exceed 2.5, and the area should not be greater than 45 m^2.

The calculated thicknesses of the reinforced concrete, RCC and continuous reinforced concrete surface layers can be determined according to the traffic load grade, highway grade and variation level grade, referring to the calculation method of ordinary cement concrete pavement.

The steel fiber volume ratio of steel fiber concrete should be 0.6–1.0%, and the surface thickness should be 0.65–0.75 times the ordinary mixed soil surface thickness, which is determined by the steel fiber content. Under heavy or heavy traffic loads, the minimum thickness should be 180 mm; under medium or light traffic loads, the minimum thickness should be 160 mm.

The thickness of the upper layer of the asphalt concrete of the compound pavement should not be less than 40 mm. The thickness of the layer under the cement mixed soil is calculated according to the method of ordinary cement concrete pavement. A bond layer should be set between the lower layer of cement concrete and the upper layer of asphalt concrete.

The thickness of the concrete surface slab depends on the highway and traffic load grade. The thickness required for ordinary concrete, reinforced concrete, RCC or continuous reinforced concrete surface slabs can be selected by referring to the range listed in Table 4.7.

To ensure the safety of driving, the surface structure of a concrete pavement slab should be constructed via the methods of grooving, pressing, grooving, drawing, or drawing. The construction depth should meet the requirements of Table 4.8 at the beginning of use.

Table 4.7 Reference range for the surface thickness of cement concrete

Traffic load class	Highway grade	Level of variation	Surface thickness (mm)
Extremely heavy	–	Low	≥ 320
Extra heavy	Highway	Low	320–280
	First class road	medium	300–260
		Low	280–240
	Second class roads	Medium	
Heavy	Highway	Low	
	First class road	Medium	270–230
		Low	260–220
	Second-class roads	Medium	
Medium	Second class roads	High	250–220
		Medium	240–210
	Third and fourth class highways	High	
		Medium	230–200
Light	Third and fourth class highways	High	220–190
		Medium	210–180

Note When asphalt concrete layers are installed on the cement concrete slab, the thickness of the slab can be reduced by 1 cm for each additional 4 cm asphalt concrete layer

Table 4.8 Requirements for surface construction depth of cement concrete surface layers of highways at all levels (mm)

Type	Highway, general highway	The secondary, third, fourth class highways
General road	0.70–1.10	0.50–1.0
Special sections	0.80–1.20	0.60–1.10

Note 1. Special sections include expressways and township highways, interchanges, level crossings or variable speed lanes, etc. Other classes of highways include sharp bends, steep slopes, intersections or areas near market towns. 2. For areas with annual rainfall halos below 600 mm, the values listed in the table can be appropriately reduced

Cement Concrete Pavement Base

The base of cement concrete pavement should have sufficient anti-erosion ability and a certain stiffness. For wet and rainy areas, the subgrade is composed of low-permeability fine-grained soil from the expressway and the first road or experiences heavy traffic or heavy traffic from the second road, and a drainage base is appropriate. The selection of the cement concrete pavement base material is shown in Table 4.9, and the appropriate thickness range of each base material is shown in Table 4.9.

The width of the base should be 300–650 mm wider than each side of the concrete panel. The shoulder adopts a concrete surface layer, the thickness of which is the

Table 4.9 Range of suitable thickness for each type of base

Types of materials		Suitable construction layer thickness (mm)
Lean concrete, RCC		120–200
The inorganic binder stabilizes the pellet		150–200
Asphalt concrete	The nominal maximum particle size of aggregate is 9.5 mm	25–40
	The nominal maximum particle size of aggregate is 13.2 mm	35–65
	The nominal maximum particle size of aggregate is 16 mm	40–70
	The nominal maximum particle size of aggregate is 19 mm	50–75
Asphalt stabilized crushed stone	The nominal maximum particle size of aggregate is 19 mm	
	The nominal maximum particle size of aggregate is 26.5 mm	75–100
Porous cement stabilized crushed stone		100–150
Graded gravel, unsieved gravel, graded gravel or crushed gravel		100–200

same as that of the roadway surface layer slab, and the width of the base should be the same as that of the roadbed.

When roller-compacted concrete(RCC) is used as the base, the longitudinal and transverse joints corresponding to the concrete pavement slab should be set. When lean concrete is used as the base, if the flexural tensile strength exceeds 1.5 MPa, the transverse joint corresponding to the concrete pavement slab should be set; if the width of sand spreading is greater than 7.5 m, longitudinal contraction joints should also be set.

Under a road surface bearing extremely heavy, extra heavy or heavy traffic loads, the subbase should be set up under the base; under medium or light traffic loads, the subbase should not be set. When inorganic binder stabilization materials are used at the base and the road bed is composed of fine-grained soil, a granular subbase should be set under the base.

The asphalt concrete interlayer should be laid on lean concrete or the RCC base, and it's thickness should not be less than 40 mm. A sealing layer should be set on inorganic binder-stabilized macadam, and the sealing layer can be treated with a single-layer asphalt surface or an appropriate membrane. When single-layer asphalt surface treatment is used, the layer thickness should not be less than 6 mm.

In rainy areas, a drainage base composed of graded asphalt-stabilized gravel or graded cement-stabilized gravel is appropriate for highways and first-class highways built on low-permeability fine-grained soil or second-class highways bearing extremely heavy or extra heavy traffic loads. The drainage base should be set up under the impervious base composed of cement-stabilized gravel, and the top surface of the bottom base should be paved with an asphalt sealing layer or waterproof geotextiles composed of an impervious bottom base.

The suitable compaction thickness of various base and base structure layers should be determined according to the requirements of the nominal maximum particle size and compaction effect of the selected aggregate. When the design thickness of the base or bottom layer exceeds the appropriate compaction thickness range of the corresponding material, the thickness of the material and other layers should be adjusted within the appropriate range first.

The designed thickness of the lean concrete or RCC base should be rounded to 10 mm according to the calculated thickness.

The calculated thickness of the drainage base of the open-graded asphalt-stabilized macadam or cement-stabilized macadam should meet the requirements for the design infiltration of surface water. The design thickness of the drainage base should be rounded to 10 mm and then increased by 20 mm according to the calculated thickness.

Subgrade and Functional Layer of Cement Concrete Pavement

The functional layer structure of cement concrete pavement is usually set up to meet the special needs of the roadbed and can be divided into three categories: an antifreeze layer, a drainage layer and a reinforcement layer.

When the total thickness of the cement concrete pavement cannot meet the minimum antifreeze thickness requirement in the seasonally frozen area, an antifreeze layer should be set to ensure that the total thickness meets the minimum antifreeze thickness requirement.

For soil subgrades with poor hydrogeological conditions, a drainage layer should be set up to prevent erosion of the pavement structure by groundwater when the humidity of the road bed soil is high.

When the subgrade soil is weak, uneven settlement and deformation may still occur after reinforcement, and a reinforcement layer should be set to enhance the bearing capacity of the road bed.

Sometimes, the above three conditions are both conditions, and when selecting functional layer structure materials, we should also consider a variety of functions. In general, the functional layer is constructed with local cheap materials, or local materials mixed with less winnowing inorganic binder, such as sand, sand and gravel, and low-dose inorganic binder, are used after treatment. The functional layer thickness is generally 150–200 mm, and the cement concrete pavement subgrade should meet the requirements of stability, dense, homogeneous, and durable to provide uniform support for the pavement structure. Therefore, the subgrade soil quality requirements are very strict. Generally, high-liquid limit clay and fine soil containing organic matter cannot be used for road bed filler on highways or highways, nor can it be used for road bed filler on secondary highways or below secondary highways. A low liquid limit viscosity with a high liquid limit powder and a plastic index greater than 16 or an expansion rate greater than 3% cannot be used for road packing on highways and township roads. When the soil must be used as filler due to the limitation of conditions, inorganic binders such as lime or cement should be added for treatment.

The design elevation of embankments should be increased for sections with high underground water levels. If the design elevation is limited and the roadbed cannot reach the critical height of the medium wet state, coarse-grained soil or low agent batch lime or cement-stabilized fine-grained material should be selected as the road bed filler, which fails to reach moist conditions.

When the critical height of the roadbed is reached, in addition to the above filling, measures to reduce the groundwater level should also be taken, such as setting drainage seepage ditches. The compactness degree of the roadbed should meet the requirements of the code for the Design of Highway Roadbed (JTG D30). The levelling layer should be laid on the top surface of the rock or rock-filled roadbed. The levelling layer can be made of unscreened gravel, stone chips or low-dose cement-stabilized granular material, and its thickness depends on the unevenness of the top surface of the roadbed, which is generally 100–150 mm.

According to the Code for Design of Highway Cement Concrete Pavement (JTG D40), the comprehensive elastic dispersion of the top surface of the road bed should not be less than 40 MPa, should not be less than 60 MPa under medium or heavy traffic loads, and should not be less than 80 MPa under extra or extremely heavy traffic.

4.2.2 Cement Concrete Pavement Thickness Design

Design Calculation Model and Selection

The regression formula in the specification is established via the finite element calculation method, and the following scheme is used for structural analysis: (1) the plane sizes of the base layer and the surface layer plate cannot be equal; (2) the finite element method is used to solve the base layer and the surface layer plate, and the load stress is the cubic elastic element. The horizontal and vertical displacements between the layers are smooth and continuous, but the layers do not experience tension. (3) The temperature warping stress is solved via an approximate analytical method. The base plate and surface plate are assumed to be thin plates, and the layers are connected by vertical linear springs.

A comparison between the different models and the assumed calculation results reveals that the elastic foundation slab theory should be adopted in the structural analysis of cement concrete pavement. In addition to the granular base, other types of base and concrete surfaces should be analyzed according to the separated double slab model. The granular base and all kinds of bottom base and functional layers should be regarded as multilayer elastic foundations together with the subgrade, and the corresponding elastic modulus of the top surface of the foundation should be used as the characteristic.

According to the different types and combinations of base and surface layers, the following mechanical models can be used for pavement structure analysis:

(1) Elastic foundation single-layer plate model—suitable for a cement concrete surface on a granular base or an old asphalt pavement with a cement concrete surface; the parts below the bottom surface of the surface plate are treated as elastic foundations.

(2) The elastic foundation double-layer plate model is suitable for inorganic binder base or asphalt base cement concrete surface layers and is based on the separated cement mixed soil layer. The surface layer and a layer or the old and new surface layers constitute a double plate, with the base layer under the bottom surface or the old surface layer under the bottom surface of the part according to the elastic foundation treatment.

(3) Composite plate model—This model is suitable for surface layer or base layer composite plates composed of two layers of different performance materials. The old cement concrete pavement is paved with cement concrete above the layer, two layers of different performance materials composed of an interlayer bonding surface layer, as a single layer of elastic foundation or elastic foundation on the upper layer of the double plate; the inorganic binder base or asphalt base and inorganic binder base composed of the base, as the lower plate of the double plate on the elastic foundation.

The critical load position of the concrete surface slab is always in the middle of the longitudinal seam edge. The critical load position of the base plate is the same as that of the surface plate.

The general steps of the design calculation are as follows:

(1) Determine the traffic volume-related parameters according to research or prediction, calculate Ne, determine the variation level, and determine the reliability coefficient.

(2) List all known conditions. According to the combination form selected by the structural combination design, the thickness of each layer and the elastic modulus of the material except the layer to be designed (generally the uppermost plate) are preset, and the modulus of the layer to be designed, Poisson's ratio and standard values of flexural strength are preset.

(3) Determine the design width of the board, the paving and transverse connection construction scheme, and the plane size of the preset board (length and width) according to the requirements of the type of selected joints.

(4) The basic model of the design calculation is selected according to the structure combination, which can be divided into three cases: a single-layer plate on an elastic foundation, a separated double-layer plate and a combined double-layer plate (or composite plate).

(5) The elastic modulus (E_0) of the subgrade is determined according to the soil quality and other clear conditions of the subgrade, and the equivalent elastic modulus (E_t) of the top surface of the foundation is calculated by combining the thickness and modulus of other layers of the elastic layered foundation.

(6) Different regression formulas are selected according to the model. The load stresses σ_{ps} and σ_{pm} generated by the standard axial load (or design axial load) and the heaviest axial load acting on the four sides of the free plate at the critical load position are calculated, and the correction coefficients K_r, K_c, and K_f are determined to calculate the load fatigue stress σ_{pr} and the maximum load stress $\sigma_{p,max}$ (only the first two coefficients are corrected).

(7) Select different regression formulas according to the model, calculate the maximum temperature stress $\sigma_{t,\,max}$ under the combined action of temperature internal stress and warping stress, determine the correction coefficient K_t, and calculate the temperature fatigue stress σ_{tr}.

(8) For the separated double-layer plate, the load fatigue stresses σ_{bps} and σ_{bpr} of the lower plate (or base) are calculated and tested in a similar way. (Note that the calculation formula of K_f is different when the materials of the upper and lower plates are different) and checks whether it meets the limit state expression of the base layer (the maximum load stress generated by the heaviest axial load is not considered, and the temperature fatigue stress term is ignored in the fatigue stress analysis).

(9) Check whether the limit state expression is established. If not, go back to step (2) and change the preset layer thickness or redesign the combination design until it is established.

(10) Reduce the thickness of the layer to be designed or choose a combination scheme with a lower total price of other materials until it just meets the limit state expression and determines the optimal design scheme.

(11) After the thickness of the calculated layer is determined, a 6 mm wear thickness should be added and rounded according to 10 mm as the design thickness of the concrete surface layer.

Comprehensive Elastic Modulus of the Elastic Foundation

The foundation model and analysis method selected by all the models in the standard design method are basically the same and are characterized by the maximum Et (also known as the equivalent elastic modulus) of the comprehensive v modulus of the top surface of the foundation. However, it should be noted that the "elastic layered foundation" differs according to the combination of the base layer and the number of layers of the panel, which should be distinguished:

(1) Under the single-layer cement concrete pavement slab, when the granular material is used as the base, the granular layer and the following layer are regarded as the foundation, including the granular layer itself.

(2) Under the single-layer cement concrete pavement plate, when the nongranular layer is the base, the layers below the base are regarded as the foundation, excluding the base itself.

(3) For the old cement concrete board, when the separated new concrete board is added, the top surface of the old concrete board is the foundation (including the base itself).

(4) When the old cement concrete plate is added and combined with the new cement concrete plate, the new and old plates are converted into a single layer according to the combined double-layer plate theory. When the old base is nongranular material, the old base is converted into a single layer according to the theory of a separated double-layer plate. The lower part is regarded as the foundation, and the whole is calculated according to the separated double-layer plate.

(5) When cement concrete pavement board is added to the old asphalt pavement, the equivalent elastic modulus is converted by the index tested on the top of the old pavement. In short, except for the single-layer slab structure with a granular base or when the old asphalt pavement is paved with a concrete face plate, the layers below the face plate belong to the foundation, and the other structures are part of the foundation after the upper nongranular material layer is removed. The equivalent elastic modulus is given in the form of the top surface index of the foundation. The same regression formula was used for different models.

$$E_t = \left(\frac{E_x}{E_0} \right)^{\alpha} E_0 \tag{4.40}$$

$$\alpha = 0.86 + 0.26 \ln h_x \tag{4.41}$$

$$E_x = \frac{\sum_{i=1}^{n} E_i h_i^2}{\sum_{i=1}^{n} h_i^2} \tag{4.42}$$

$$h_x = \sum_{i=1}^{n} h_i \tag{4.43}$$

where

E_0 Comprehensive elastic modulus of the subgrade top surface (MPa);

α The regression coefficient related to the total thickness h_x of each layer in the foundation except the subgrade;

E_t Equivalent elastic modulus of the top surface of the foundation (MPa);

h_x The total thickness of each layer in the foundation except the subgrade (m);

n Number of layers of elastic foundation (excluding the subgrade half-space body);

E_i, h_i elastic modulus (MPa) and thickness (m) of the ith structural layer;

E_x Equivalent elastic modulus of the granular layer (MPa).

When the cement concrete surface is paved on the old asphalt concrete pavement, the maximum elastic modulus of the top surface of the foundation can be calculated according to the bending and settling determination results of the center point of the drop weight bending apparatus (load 50 kN, bearing plate diameter 30 cm) according to Eq. (4.44):

$$E_t = \frac{18621}{\omega_0} \tag{4.44}$$

Alternatively, the bending and settling determination results of the Beckman beam (vehicle loading with rear axle weight of 100 kN) can be calculated according to Eq. (4.45):

$$E_t = \frac{13739}{\omega_0^{1.04}}$$ (4.45)

$$\omega_0 = \overline{\omega} + 1.04 s_w$$ (4.46)

where

ω_0 The road section represents the bending value (0.01 mm);
$\overline{\omega}$ Average bending and subsidence of the road section (0.01 mm);
s_w Standard deviation of road buckling (0.01 mm).

Design Method of the Single-Layer Plate Model

(1) The load stress

1. Calculation of the load fatigue stress of a concrete slab

(a) The design axis is loaded at the critical load position of the four-sided free plate σ_{ps}

$$\sigma_{ps} = 1.47 \times 10^{-3} r^{0.70} h_c^{-2} P_S^{0.94}$$ (4.47)

$$r = 1.21 \sqrt[3]{\frac{D_c}{E_t}}$$ (4.48)

$$D_c = \frac{E_c h_c^3}{12(1 - v_c^2)}$$ (4.49)

where

P_s Design the single axle load (kN);
H_c, E_c, and v_c are the thickness (m), flexural elastic modulus (MPa) and Poisson's ratio of the mixed soil surface laminate, respectively;
r Relative stiffness radius of the concrete surface slab (m);
D_c Section bending stiffness of the concrete slab (MN· m);
E_t Equivalent elastic modulus of the slab foundation (MPa).

(b) Three correction coefficients were determined: k_r, k_c, and k_f

The stress reduction factor KR has a positive effect on the stress reduction of the plate because the load transfer capacity of the joint is less than or equal to 1. Since the critical load position is at the edge of the longitudinal seam, it is mainly determined by the condition of the road shoulder: 0.87–0.92 when the concrete shoulder is used (the low value is taken when the shoulder layer is equal to the road surface layer, and the high value is taken when the thickness is reduced); 1 is used when the flexible shoulder or dirt shoulder.

Considering the difference between theory and practice and the influence of the dynamic load and other factors, the comprehensive coefficient KC is determined according to highway grade Table 4.10.

The load fatigue stress coefficient K_t is related to the cumulative axial order N_e and can be determined via Eq. (4.50):

$$K_\mathrm{f} = N_e^\lambda \tag{4.50}$$

where

N_e Cumulative number of actions of the design shaft load during the design reference period;

λ Material fatigue index: 0.057 for ordinary concrete, reinforced concrete and continuous reinforced concrete; 0.065 for RCC and lean concrete; steel fiber-reinforced concrete is calculated according to Eq. (4.51).

$$\lambda = 0.053 - 0.017\rho_\mathrm{f}\frac{l_f}{d_f} \tag{4.51}$$

where

ρ_f Volume ratio of steel fibers (%);
l_r The length of a steel fiber (mm);
d^f Diameter of the steel fiber (mm).

(c) Load fatigue stress calculation

$$\sigma_\mathrm{pr} = K_R K_C K_\mathrm{f}\sigma_\mathrm{ps} \tag{4.52}$$

2. Load stress calculation of surface laminates under the heaviest axial load.

(a) The maximum axial load (or ultimate load) is the load stress σ_pm at the critical load position of the four-sided free plate.

The formula is the same as σ_ps, but the standard (or design) shaft load p_s is replaced by the heaviest shaft load P_m in the formula.

Table 4.10 Comprehensive coefficient K_C

Highway grade	The highway	First class road	The secondary roads	Third and fourth class highways
K_c	1.15	1.10	1.05	1.00

(b) Determining the correction factors K_r and K_C

The determination methods of K_r and K_C are the same as those used for the calculation of the load fatigue stress, without repeated calculations.

(c) The maximum load stress of the heaviest axle at the critical load position $\sigma_{p,max}$

$$\sigma_{P,max} = K_R K_C \sigma_{pm} \tag{4.53}$$

(2) Temperature stress

The temperature stress is not directly related to whether the load is a repeated load or a single heaviest load. However, when the load stress is added to the temperature stress, there is a similarity problem with the real state. At the beginning of the use period of the road surface, owing to the strong foundation constraint, the temperature shrinkage and warning of internal stress are high. In the later stage, under the repeated action of stress, the constraint on the interface weakens, so the temperature fatigue stress decreases. When considering the effect of fatigue, the load fatigue stress should be used, and the temperature fatigue stress should also be used. In the investigation of the action of the heaviest axial load, the maximum temperature stress is chosen without considering the fatigue effect because its action is one-off.

1. Maximum temperature stress of surface laminates $\sigma_{t,max}$

(a) Temperature stress coefficient combining the temperature warping stress and internal stress B_L

$$B_L = 1.77e^{-4.48H_c} C_L - 0.131(1 - C_L) \tag{4.54}$$

$$C_L = 1 - \frac{\sinh t \times \cos t + \cosh t \times \sin t}{\cos t \times \sin t + \sinh t \times \cosh t} \tag{4.55}$$

$$t = \frac{L}{3R} \tag{4.56}$$

where

C_L Temperature warping stress coefficient of the concrete slab;
L Transverse seam spacing of surface laminates, that is, plate length (m);
r Relative stiffness radius of the surface laminates (M).

(b) Maximum temperature stress

$$\sigma_{t,max} = \frac{\alpha_c E_C H_C T_g}{2} B_L \tag{4.57}$$

where

α_c Linear expansion coefficient of the concrete;

T_g Maximum temperature gradient at the highway site once every 50 years.

2. Temperature fatigue stress of surface laminates σ_{tr}

(a) Determine the temperature fatigue stress coefficient k_t

$$K_t = \frac{f_r}{\sigma_{t,\max}}\left[a_t\left(\frac{\sigma_{t,\max}}{f_r}\right)^{b_t} - c_t\right]$$

(4.58)

where

a_t, b_t, c_t The regression coefficient.

(b) Temperature fatigue stress σ_{tr}

$$\sigma_{tr} = k_t\sigma_{t,\max}$$

(4.59)

Design Method of the Separated Double-Layer Plate Model

When RCC or lean concrete is used as a base, it is necessary to check whether the load fatigue stress of the base exceeds the flexural and tensile strengths. When other materials are used as the base layer, compared with the theory of a single-layer plate on an elastic foundation mentioned above, although the influence of the large stiffness of the base layer is considered in the calculation formula, the limit state of the base layer is not considered, so it is not necessary to calculate the stresses of the base layer. Attention should be given to the selection of the formula for practical calculation.

(1) Load stress

1. Load stress caused by the critical load position of the load acting on four free plates

The load fatigue stress calculation of the upper plate is similar to that of the single-layer plate model, but it is different from the calculation formula of the load stress under the P_s action of the designed axial load.

(a) Load stress of the upper plate under the design load

$$\sigma_{ps} = \frac{1.45 \times 10^{-3}}{1 + \frac{D_b}{D_c}}r_g^{0.65}h_c^{-2}P_s^{0.94}$$

(4.60)

$$D_b = \frac{E_b h_b^3}{12\left(1 - v_b^2\right)}$$

(4.61)

$$r_g = 1.21 \sqrt[3]{\frac{D_c + D_b}{E_t}} \tag{4.62}$$

where

D_b	Section bending stiffness of the lower plate (MN·m);
H_b, E_b, v_b	thickness of the lower plate (m), modulus of the flexural elastic piece (MPa) and Poisson's ratio;
r_g	Total relative stiffness radius of the double slab (M);
h_c, D_c	Thickness of the upper plate (m) and bending stiffness of the section (MN·m).

(b) Load stress of the upper plate under the heaviest axial load

The formula used is the same as above, but the heaviest axle load P_m is used instead of the design axle load P_s.

(c) Load stress of the lower plate under the design load

The load stress of the RCC, lean concrete or cement concrete subplate needs to be calculated:

$$\sigma_{bps} = \frac{1.41 \times 10^{-3}}{1 + \frac{D_c}{D_b}} r_g^{0.68} h_b^{-2} P_s^{0.94} \tag{4.63}$$

2. Load fatigue stress and maximum load stress

(a) Correction factors of the load fatigue stress k_r, k_c, and k_f

Upper plate: the determination of three correction coefficients is the same as that of a single-layer plate on an elastic foundation, and the formula for calculating the load fatigue stress is the same.

(b) Maximum load stress of the upper plate under the heaviest axial load

The method of determining the two correction coefficients K_r and K_C of the upper plate is the same as that of the single-layer plate of the elastic foundation, and the calculation formula of the maximum load stress is the same.

(c) The lower plate considers the load fatigue stress with the correction factor:

$$\sigma_{bpr} = k_c k_f \sigma_{ps} \tag{4.64}$$

(2) Temperature stress

The temperature stress of the lower plate is not considered.

(a) Maximum temperature warping stress of the upper plate

Compared with the single-layer slab model on an elastic foundation, except for the calculation formula of the temperature warping stress coefficient C_L, the other calculation formulas are the same:

$$C_L = 1 - \left(\frac{1}{1+\xi}\right)\frac{\sinh t \cdot \cos t + \cosh t \cdot \sin t}{\cos t \cdot \sin t + \sinh t \cdot \cosh t} \tag{4.65}$$

$$t = \frac{L}{3r_g} \tag{4.66}$$

$$\xi = -\frac{\left(k_n r_g^4 - D_c\right)r_\beta^3}{\left(k_n r_\beta^4 - D_c\right)r_g^3} \tag{4.67}$$

$$r_\beta = \sqrt[4]{\frac{D_c D_b}{(D_c + D_b)k_n}} \tag{4.68}$$

$$k_n = \frac{1}{2}\left(\frac{h_c}{E_c} + \frac{h_b}{E_b}\right)^{-1} \tag{4.69}$$

where

ξ Parameters related to double slab construction;

r_β Interlayer contact condition parameters (m);

k_n When there is no asphalt concrete interlayer or isolation layer between the upper and lower layers, the vertical contact stiffness between the surface layer and the base layer is calculated according to (4.68), and the isolation layer is set to 3000 MPa/m.

(b) Temperature fatigue stress of the upper plate

In the calculation of the temperature fatigue stress, the other correction coefficients and formulas are the same as those of the single-layer plate model on an elastic foundation, except for C_L and σ_t, max.

Composite Panel Model Design Method

The surface layer composite plate can be treated as one layer, which is analyzed on the basis of the calculation method of single-layer plate design. However, its D_C and h_C indices should be modified, and other calculation parameters and formulas should be calculated with $\widetilde{D}_c, \widetilde{h}_c$. The modifications are as follows:

$$\tilde{D}_c = \frac{E_{c1}h_{c1}^3 + E_{c2}h_{c2}^3}{12(1 - v_{c2}^2)} + \frac{(h_{c1} + h_{c2})^2}{4(1 - v_{c2}^2)}\left(\frac{1}{E_{c1}h_{c1}} + \frac{1}{E_{c2}h_{c2}}\right)^{-1} \tag{4.70}$$

$$\tilde{h}_c = 2.42\sqrt{\frac{D_c}{E_{c2}d_x}} \tag{4.71}$$

$$d_x = \frac{1}{2}\left[h_{c2} + \frac{E_{c1}h_{c1}(h_{c1} + h_{c2})}{E_{c1}h_{c1} + E_{c2}h_{c2}}\right] \tag{4.72}$$

where

E_{c1}, h_{c1} Flexural elastic modulus of the upper layer of the surface composite plate (MPa) and thickness (m);

E_{c2}, v_{c2}, and h_{c2} Flexural elastic modulus of the lower layer of the surface composite plate (MPa), Poisson's ratio and thickness (m);

d_x Distance from the neutral axis of the surface composite plate to the bottom of the lower layer (m).

(1) Maximum temperature stress correction of the surface composite plate

$$\sigma_{t,\max} = \frac{\alpha_c T_g E_{c2}(h_{c1} + h_{c2})}{2}B_L\zeta \tag{4.73}$$

$$\zeta = 1.77 - 0.27\ln\left(\frac{h_{c1}E_{c1}}{h_{c2}E_{c2}} + 18\frac{E_{c1}}{E_{c2}} - 2\frac{h_{c1}}{h_{c2}}\right) \tag{4.74}$$

where

B_L The calculation method of the temperature stress coefficient of the surface layer composite plate is the same as that of the single-layer plate model. The thickness of the surface layer plate, HE, is the total thickness of the composite plate ($h_{c1} + h_{c2}$), and the temperature warping stress coefficient, CL, is calculated according to the single-layer plate model formula for the single-layer plate and according to the separated double-layer plate model formula for the double-layer plate.

ξ Maximum temperature stress correction coefficient of the surface composite plate.

1. Base composite plate

When the base is a composite plate, it is equivalent to three rigid layers, which is similar to the situation in which the RCC or lean concrete base is replaced by a combined double-layer plate. Before the separated double-layer plate model is applied, the bending stiffness of the base (composite plate) should be modified:

$$D_{b0} = D_{b1} + D_{b2} \tag{4.75}$$

$$\sigma_{bps} = \frac{\tilde{\sigma}_{bps}}{1 + \frac{D_{b2}}{D_{b1}}} \tag{4.76}$$

where

D_{b0} Bending stiffness of the base composite plate (MN·m);

D_{b1}, D_{b2} The bending stiffnesses (MN·m) of the base and the other base are calculated according to the thicknesses h_{b1} and h_{b2} of the base and the other base and the elastic moduli E_{b1} and E_{b2}, respectively.

$\tilde{\sigma}_{bps}$ The nominal load stress of the base composite plate is calculated according to the separated double layer plate, where the base thickness h_{b1} is used to replace the middle base thickness h_b, and the bending stiffness D_{b0} of the composite plate is used to replace the bending stiffness D_b of the middle base plate.

The bending stiffness of the base layer composite plate is replaced by the bending stiffness of the base layer in the calculation formula of the separated double-layer plate model, and the load stress and temperature stress of the double-layer plate are calculated.

When the base is lean concrete or RCC, the load fatigue stress σ_{bpr} of the base of the composite plate should be calculated according to Eq. (4.76), and other types of bases do not need to calculate the load fatigue stress.

4.3 Mechanistic–Empirical Pavement Design

The Mechanistic-Empirical (M-E) pavement design methodology integrates physical principles with empirical observations to predict pavement performance. Critical structural responses—including stresses, strains, and deflections—are governed by traffic loads, environmental conditions, and the inherent mechanical properties of pavement materials. These relationships are mathematically modeled to quantify pavement behavior under service conditions. To bridge theoretical predictions with real-world performance, empirical distress models (e.g., fatigue cracking and rutting) are calibrated against field data to establish failure thresholds. By synthesizing mechanistic simulations with empirical validation, the cumulative effects of repeated loading can be rigorously evaluated. This hybrid framework ensures a robust linkage between material properties, structural responses, and observed pavement deterioration, forming the scientific foundation of M-E design [14].

4.3.1 M-E Pavement Design Framework

The M-E pavement design process is structured into three iterative phases to systematically integrate theoretical modeling with empirical validation [14].

Preliminary Evaluation and Input Development

This phase is initiated by establishing feasible design strategies aligned with project objectives. Subsequently, the compilation of critical input parameters is conducted, including material properties (e.g., elastic moduli and thermal coefficients), traffic characterization data (axle-load spectra, traffic growth rates, and vehicle class distributions), and climate variables (temperature gradients, precipitation patterns, and freeze–thaw cycles). These inputs form the foundational dataset for subsequent analyses.

Structural Analysis and Performance Prediction

Phase 2 employs mechanistic models to simulate the structural responses of trial pavement sections under operational conditions. Cumulative damage is quantified through empirical transfer functions, such as fatigue life equations, whereas long-term distress evolution (e.g., crack propagation and surface roughness) is predicted via incremental damage summation. The design is iteratively refined through feedback loops until predicted performance metrics-such as fatigue life and rutting depth-converge with predefined failure criteria, exemplified by AASHTO serviceability thresholds.

Life Cycle Optimization

Phase 3 extends beyond structural adequacy to evaluate economic and environmental viability. Structurally feasible alternatives undergo comprehensive cost–benefit analyses, integrating sustainability metrics (e.g., carbon footprint reduction and material recyclability) and risk assessments to identify optimal solutions. This phase ensures that the final design balances technical performance, lifecycle costs, and environmental stewardship, thereby aligning with modern infrastructure resilience standards.

4.3.2 Hierarchical Input System

The Mechanistic-Empirical (M-E) framework implements a tiered input hierarchy to optimize the trade-off between data precision and practical feasibility in pavement design. Level 1 relies on direct field measurements and laboratory-tested material properties, such as site-specific traffic monitoring and asphalt binder rheology. These inputs are required for high-volume transportation corridors (e.g., interstate highways or urban arterials), where prediction inaccuracies could lead to premature structural failures with significant economic and safety repercussions.

Level 2 utilizes data synthesized from regional agency databases, including historical axle-load distributions and climate zone averages, to reflect localized operational patterns. This tier serves as a pragmatic compromise for regional road networks, enabling designers to leverage empirical trends while avoiding the costs of exhaustive site-specific data collection.

Level 3 adopts conservative estimates from national standards (e.g., AASHTO climatic zone classifications) or empirically validated correlations. These generalized inputs, although less precise, provide a cost-efficient solution for low-risk applications such as rural access roads or parking facilities, where traffic volumes and performance demands are minimal.

This hierarchical structure systematically aligns input fidelity with project criticality, ensuring that resource allocation is proportional to the consequences of design uncertainty.

4.3.3   Environmental Effects

Climate factors (e.g., temperature, moisture, freezing index, etc.) may significantly impact road structures and subgrades, affecting their stability and strength, causing damage, and ultimately reducing their service life. Therefore, a program that integrates heat and moisture flow and simulates changes in the material and structural performance of pavement and subgrades under climate conditions is necessary.

4.3.4   Traffic Loading Data Requirements

Instead of using an equivalent single axle load to represent traffic flow over the whole life cycle, full axle-load spectrum traffic data are more appropriate for determining the axle load applied to the pavement with cumulative damage over time.

As mentioned above, the specific traffic data required are as follows:

Truck traffic volume in the base year and estimated growth rate in design life;
Hourly and monthly traffic volume adjustment factors for trucks;
Load spectra and vehicle class distribution;
The general traffic data include axle configuration, wheelbase and axles per truck.

4.3.5   Material Parameters

The material parameters in the M-E framework necessitate three fundamental parameter categories to quantify the pavement behavior under multifactorial stresses.

Mechanical Properties

They form the basis of structural response modeling, encompassing both linear elastic parameters (e.g., layer-specific elastic moduli E_1, E_2, E_3 and Poisson's ratios *v*) for stress–strain analysis and viscoelastic properties such as creep compliance to capture asphalt's time-dependent deformation under sustained loading.

Damage Resistance Metrics

They are critical for predicting long-term performance degradation, including the fatigue endurance limits of asphalt binders to quantify cracking thresholds, rutting susceptibility evaluated through flow number tests (AASHTO T 378), and fracture toughness determined via semicircular bend tests (ASTM D8044) to assess crack propagation resistance.

Environmental Sensitivity Parameters

They address climate-induced deterioration mechanisms: thermal expansion coefficients quantify dimensional instability under temperature fluctuations, whereas moisture susceptibility, tested via the modified Lottman procedure (AASHTO T 283), can be used to evaluate strength loss due to water infiltration. This tripartite characterization ensures comprehensive integration of material physics, damage mechanics, and environmental interactions within the M-E analytical framework.

Discussion, Questions and Exercises

(1) Briefly describe the three main parts of the asphalt pavement design, and explain their respective functions.
(2) What are the design indicators and standards of asphalt pavement? The corresponding asphalt pavement damage modes are analyzed.
(3) What are the current design standards for asphalt pavement in China? Trial access to information of the current standards about reflection of the actual mechanical conditions of the road surface is needed to analyze what deficiencies are present and need to be improved.
(4) Compared with asphalt pavement, what are the characteristics of the theory and model used in cement pavement structure design? Please explain and analyze it according to the characteristics of cement concrete pavement.
(5) What models are included in the elastic foundation plate theory used in the structural analysis of cement concrete pavement, and what conditions are they applicable? Combined with the design method of each model, the differences between the calculation methods of load stress and temperature stress are described.

References

1. ZHAOHONG Z, ZHIHONG X. Flexible Pavement Design Theory and Method [M]. Tongji University Press, 1987
2. MONISMITH C L. ANALYTICALLY BASED ASPHALT PAVEMENT DESIGN AND REHABILITATION: THEORY TO PRACTICE, 1962–1992 [J]. Transportation Research Record, 1992
3. TURTSCHY J C. Advanced Model for Analytical Design of European Pavement Structures [J]. Construction, 2000
4. VALKERING C P, STAPEL F D. The Shell Pavement Design Method on a Personal Computer [C]. Proceedings of the International Conference on Asphalt Pavements, Nottingham: ISAP, F, 1992
5. ZUKANG Y. Highway Design Manual: Pvament [M]. China Communications Press, 1993
6. AMERICANASSOCIATIONOFSTATEHIGHWAY, TRANSPORTATIONOFFICIALS. AASHTO Guide for Design of Pavement Structures, 1993 [M]. AASHTO guide for design of pavement structures, 1993
7. FINN F N, MONISMITH C L. Asphalt Overlay Design Procedures [J]. Nchrp Synthesis of Highway Practice, 1984
8. LIJUN S. Structural Behavior of Asphalt Pavement [M]. Tongji University Press, 2013
9. SHOOK J F, FINN F N, WITCZAK M W, et al. Thickness Design of Asphalt Pavements - the Asphalt Institute Method [C]. Proceedings of the Proceedings of 7th International Conference on Asphalt Pavements, Delft: ISAP, 1982(1): 17–44
10. JABLONSKI B, REGEHR J, REMPEL G. Guide For Mechanistic-Empirical Design of New and Rehabilitated Pavement Structures [J]. Final Report Part Design Analysis, 2001
11. WEITIAN Z, QUN Y, WENLIANG W, et al. Structural Condition Assessment and Fatigue Stress Analysis of Cement Concrete Pavement Based on the GPR and FWD [J]. Construction and Building Materials, 2022, 328
12. AVISHRESHTH, CHOPRA T. Design Aspects of Modern Cement Concrete Pavements in India [J]. International Journal of Earth Sciences and Engineering, 2017, 10(2):455–461
13. LIJUN S, HAOGAO B, ZUKANG Y. A new Method for Inverse Calculation of Slab Modulus of Cement Concrete Pavement - inert Bending Method [J]. China Civil Engineering Journal, 2000, 33(01): 83–7+99
14. Danny X.XIAO, Kelvin C.P.WANG, Kevin D.HALL.MECHANISTIC-EMPIRICAL PAVEMENT DESIGN GUIDE(MEPDG):A BIRD'S-EYE VIEW[J].Journal of Modern Transportation, 2011, (2): 114–133

Chapter 5
Pavement and Subgrade Drainage Design

5.1 Introduction

5.1.1 Sources of Water and Their Effects

The water flow affecting the subgrade and pavement can be classified into two categories according to the different water sources: surface water and groundwater, and the subgrade drainage engineering methods corresponding to this purpose can be divided into surface drainage and underground drainage [1].

Surface water includes precipitation (rain and snow) as well as sea, river, lake, channel and reservoir water. Surface water causes infiltration and erosion of roadbeds.

Infiltration and erosion may compromise the overall stability of the subgrade, resulting in water-related damage. The moisture infiltrating into the subgrade soil makes the soil too wet and reduces the strength of the road. At the same time, the water landing on the surface of the subgrade and pavement penetrates the interior of the pavement structure through pavement cracks, joints or surface voids, which adversely affects the pavement structure.

The groundwater includes upper stagnant water, phreatic water, interlayer water, etc. The degree of damage to the subgrade varies with different conditions. Light can make the subgrade wet and soft and reduce the strength of the subgrade. Heavy rain will cause frost heave, mud pumping or slope collapse, or even the whole subgrade sliding along the inclined base. Water may also cause destructive damage to roadbeds mixed with expansive soil. When the water table is high, the groundwater will rise into the pavement structure through the capillary action. In addition, when temporary stagnant water is present in the central divider and on both sides of the road, moisture from this water may also seep the interior of the road structure, which has adverse effects on the interior of the road structure.

The harmful effects produced in subgrade and pavement structures can be summarized as follows:

© Tongji University Press Co., Ltd. 2026

H. Li et al., *Road Engineering*, https://doi.org/10.1007/978-981-95-6659-4_5

(1) In particular, the strength of cohesive granular materials and foundation soil is reduced.
(2) Cement concrete pavement produces mud, followed by the wrong stage, cracking and damage to the entire shoulder.
(3) Due to the high hydrodynamic pressure generated by a moving vehicle, the fine particles of the asphalt pavement base produce mud, resulting in the loss of support.
(4) Water causes uneven freezing and swelling.
(5) Frequent contact with water spalls asphalt, affecting the durability of the asphalt concrete and resulting in cracking.

5.1.2 Subgrade and Pavement Drainage Design Tasks

The task of subgrade and pavement drainage is to reduce the soil moisture within the subgrade range to a specified limit. The subgrade is maintained in a dry state throughout the year to ensure that the subgrade and pavement have sufficient strength and stability. During the design of the subgrade, the ground water that affects the stability of the subgrade must be intercepted or diverted outside the range of the subgrade land, and the surface water should be prevented from overflowing, ponding or infiltrating. The groundwater affecting the stability of the subgrade and pavement should be cut off, drained or lowered, and guided to an appropriate place outside the subgrade range [2].

In the construction of subgrades and pavement, the design of the entire subgrade drainage system should be checked first to determine whether it is complete and proper, and if necessary, it should be supplemented or modified to ensure the optimal quality and performance of the drainage process. In addition, according to the actual situation and needs, temporary drainage measures should be set up at the construction site to ensure that the embankment fill materials and affiliated structures are operated under normal conditions, eliminate the subgrade base and soil body and water-related hidden dangers, ensure the quality of subgrade and pavement engineering, and improve construction efficiency.

In the maintenance of subgrades and pavement, drainage facilities should be inspected and maintained regularly to ensure the normal use of drainage facilities and smooth water flow, and the drainage conditions of the subgrade and pavement should be continuously optimized according to the actual situation.

5.1.3 Main Content of Subgrade and Pavement Drainage Design

The main contents of the subgrade and pavement drainage design include road boundary surface drainage; internal drainage of pavement; road boundary drainage;

drainage of highway structures, underpass roads and facilities along the road; and drainage in special areas and special sections. Therefore, the design process of highway drainage mainly includes the overall design of the drainage system, hydrological investigations and analyses, the structural form and material selection of drainage facilities, and the content of the water conservancy meter and pan.

Road boundary surface drainage includes the removal of road (bridge) surfaces, central dividers, slopes and surface water flowing into road boundaries from road adjoining areas or crossing roads.

The internal drainage of pavement includes the pavement edge drainage system, drainage layers or drainage cushions alone or in combination.

The underground drainage of road boundaries includes underground drainage facilities such as interceptor drains, seepage ditches, seepage wells, seepage tunnels or inverted drainage pipes, which can intercept and divert the groundwater of aquifers, reduce the groundwater level or drain the groundwater of slope bodies.

The drainage of highway structures, underpassing roads and facilities along the line includes bridge deck drainage, bridge (culvert) and retaining structure drainage, tunnel drainage, underpassing road drainage, the drainage of facilities along the line, etc.

Drainage in special areas and special road sections includes drainage in permafrost areas, expansive soil areas, loess areas, saline soil areas, landslide road sections and water environment sensitive road sections.

5.1.4 General Principles of Subgrade and Pavement Drainage Design

The general principles of subgrade and pavement drainage design are as follows:

(1) Drainage facilities should be adapted to local conditions, comprehensive planning, reasonable layouts, and comprehensive management, paying attention to cost-effectiveness and making full use of the existence and shape of the natural water system. Under normal circumstances, surface and subsurface drainage ditches should be short rather than long so that the water flow is not too concentrated, and timely evacuation and nearby diversion should be realized.

(2) All kinds of road drainage ditches should be set, and the coordination with farmland water conservancy; if necessary, can appropriately add disaster relief culverts or increase the culvert pipe aperture to prevent agricultural water from affecting the stability of the roadbed. Subgrade side ditches should generally not be used as farmland irrigation channels; when the two must be combined, the section of the side ditch should be enlarged and strengthened to prevent water flow from overflowing and permeating the subgrade.

(3) Pre-design investigations shall include: the water and geological conditions should be determined; the key link to the comprehensive planning of the drainage system should be the roadbed drainage used to match the plant layout; the

surface drainage and underground drainage of the drainage ditch layout used to match the vertical layout, and the comprehensive design and installation of the building roadbed. A road section with difficult drainage and poor geology should be combined with roadbed protection and reinforcement, and a special design should be carried out.

(4) Roadbed and pavement drainage should pay attention to preventing soil erosion and water loss near hillsides, and try not to destroy the natural water system, which is not easily combined with the natural gully stream and changes the nature of water flow, as much as possible to choose favorable geological conditions to lay artificial ditches and reduce the protection and reinforcement of drainage ditches. For the main drainage facilities in key sections, as well as the drainage ditches in soft soil and steep slope sections, we should pay attention to the necessary protection and reinforcement.

(5) The subgrade and pavement drainage should be combined with the local hydrological conditions, the road grade and other specific conditions, and attention should be given to local materials to ensure that they are not only stable and applicable but also economical.

(6) Hydraulic and hydrological calculations of highway drainage refer to the Code for Design of Highway Drainage (JTG/T D33) [3].

5.2 Pavement Drainage Design

5.2.1 Road Boundary Surface Drainage

Slope Drainage

The commonly used subgrade surface drainage facilities include side ditches, interceptor ditches, drainage ditches, drop structures and chutes, etc., and culverts, inverted siphons and retention basins when necessary. These drainage facilities are located in different parts of the subgrade, and their drainage functions, layout requirements and structural forms are different. For the runoff calculation of subgrade surface drainage facilities, 15 years should be adopted for expressways and first-class highways, and the maximum rainfall intensity of any 30 min within the return period of 10 years should be adopted for other grade highways. The top of each type of surface water trench should be more than 0.2 m above the design water level [4].

(1) Ditches

The side ditch is arranged on the lateral shoulder of the subgrade or the lateral foot of the slope of the low embankment, which is parallel to the centerline of the road. It is used to collect and exclude minor surface water within the range of the roadbed and flows to the roadbed. The roadside pit in the road section filled with flat ground is usually integrated with the drainage design of the roadbed so that it can drain the side ditch.

Given the typically minimal flow in side ditches, hydrologic and hydraulic calculations are generally not required. The standard cross-sectional form is selected according to the specific conditions along the line. Side ditch is close to the roadbed and usually do not allow the flow of water from other drainage ditches and cannot be used in conjunction with other man-made ditches.

The side ditch should not be too long, as much as possible, to wipe the trench water near the discharge to the roadside natural ditch or low-lying area; if necessary, to set up a culvert, the side ditch water across the roadbed from the other side should discharge.

The longitudinal slope of the side ditch (except near the outlet) is generally consistent with the longitudinal slope of the route. For the flat slope section, the side ditch should be maintained at a longitudinal slope of not less than 0.5%. In special cases, 0.3% can be used, but the outlet spacing of the side ditch should be reduced. In the vicinity of the exit of the side ditch and the difficult sections of drainage, such as the curve of turning back and the flat curve of the high embankments, the side ditch should be specially designed.

The transverse section of the side groove includes a trapezoid, rectangle, triangle, and flow line, as shown in Fig. 5.1. The transverse section of the side ditch is generally trapezoidal, and the slope of the inner side of the trapezoidal side ditch is 1:1.5–1:1.5. The slope of the outer side slope is the same as the slope of the excavation side slope. The side ditch of the stone section should be rectangular transverse by, the inner side slope should be upright, the slope should be protected by slurry and stone, and the slope of the outer side slope should be the same as the slope of the excavation side slope. The soil side trench of the shallow excavation section with little rain can adopt a triangular cross section, the inner side slope should adopt a ratio of 1:2–1:3, and the slope of the outer side slope should be the same as that of the excavation side slope. The flow conditions of triangular slopes are poor, so the ditch depth should be increased appropriately when the flow is high.

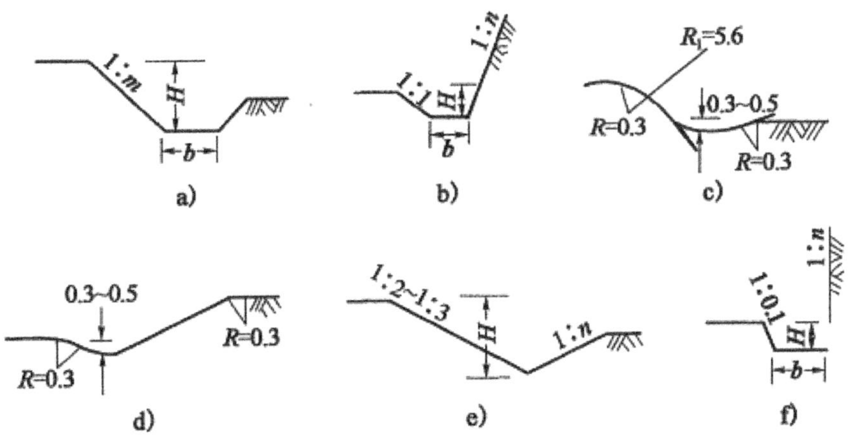

Fig. 5.1 Schematic diagram of the transverse form of the marginal sulcus

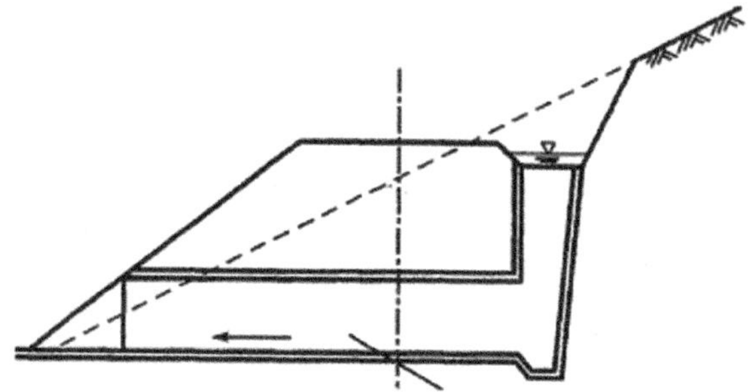

Fig. 5.2 Side ditch discharge profile of inspection wells in front of culverts

The bottom width and depth of the trapezoidal side ditch are 0.4–0.6 m. In areas or sections with minimal water flow, the lower limit is taken or smaller, but it should not be less than 0.3 m; if the precipitation is concentrated or the terrain is low, the upper limit or larger should be taken. The shoreline side ditch is the side that will cross section the embankment.

The corners are rounded to prevent sand or snow accumulation on the subgrade slope. It is suitable for subgrades in areas of sand or snow.

The side ditch can be protected by mortar shingles, mortar pebbles and precast cement concrete blocks. The mortar strength for masonry should be M7.5 for expressways and first-grade highways and M5 for other grade highways. Near the outlet of the edge ditch, the erosion of water flow is serious, so it must be carefully arranged, and corresponding measures should be taken.

When the water flow from the edge ditch flows to the inlet of bridge and culvert, proper treatment should be applied to avoid flushing the water flow from the edge ditch. Figure 5.2 shows an example of setting an inspection well at the entrance of the culvert. In addition, according to the terrain and other conditions, before the entrance of the bridge and culvert or other water flow gap is large, the jet trough and fall and other structures are set up, and the water flows into the bridge and culvert or other designated sites.

When the side ditch water flows to the reverse curve section, the general side ditch water is full, and a greater rate of flow is appropriate for setting up along the side ditch direction along the hillside diversion ditch. The water will be diverted to the subgrade range outside the natural ditch or set up rapids slot basket hole and other structures, and the water will be diverted down the hillside or subgrade on the other side so as not to cause erosion in the reverse curve section.

(2) Intercepting ditch

A cutoff ditch, also known as a trench, is generally set in the excavation of the subgrade slope outside the top or hillside embankment above the appropriate place to intercept and exclude the subgrade above the ground runoff to the subgrade and alleviate hydraulic load on side ditches to ensure that the excavation slope and fill the foot of the slope are not washed by water. A road section with less precipitation or a hard slope and a low slope so that the impact of erosion is not large cannot set up a ditch; in contrast, if a ditch is set up, if the precipitation is greater and the frequency of heavy rain is greater, the hillside cover is relatively soft, the slope is greater, and soil and water loss is more serious; if necessary, two or more ditches can be set up.

Figure 5.3 is one of the legends of the cutoff ditch set above the side slope of the cutting section. The distance d in the figure should be greater than 5.0 m in general, and 10.0 m or greater in the poor geological area. One side of the lower part of the cut ditch can pile the earthwork to dig the ditch, and the top of the ditch dip is 2% at this stage. Figure 5.4 shows the position and section size of the ditch when the mound is installed above the cut.

The cross-sectional form of a ditch is generally trapezoidal, and the slope of the ditch is determined by geotechnical conditions, generally 1:1.0–1:1.5, as shown in Fig. 5.5. The width of the bottom of the ditch should not be less than 0.5 m, and the depth of the ditch should not be less than 0.5 m according to the design flow.

The ditch should be located perpendicular to the predominant overland flow to improve the efficiency of water capture and shorten the length of the ditch. The drainage ditch should ensure smooth flow of water and discharge into the nearest natural ditch. When necessary, drainage structures such as jet troughs or culverts direct water into designated locations. The water flow in the ditch should not enter the side ditch. When it must enter, the cross section of the side ditch should be enlarged and protected.

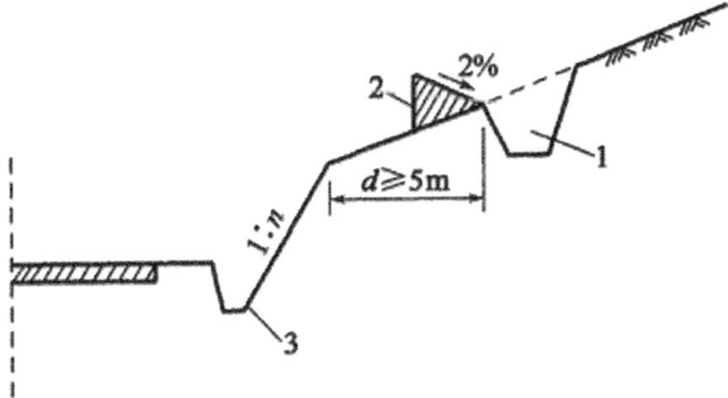

Fig. 5.3 Schematic diagram of the cutoff ditch in the excavation section

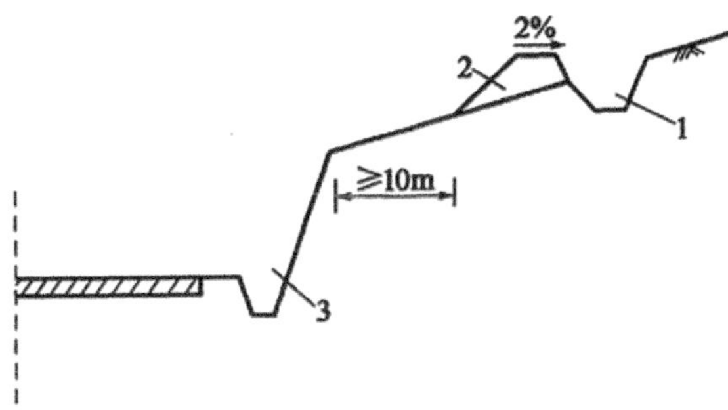

Fig. 5.4 Relation diagram of the abandoned mound and cutoff ditch in the excavation section

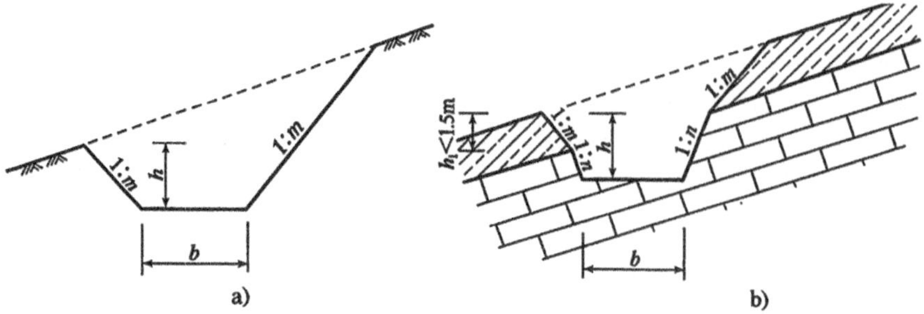

Fig. 5.5 Legend of the cross section of the gully cut

The trench bottom should have more than 0.3% longitudinal slope, and the trench bottom and walls should be flat and dense, with no stagnant flow and no seepage; if necessary, reinforcing and paving the length of the ditch to 200–500 m is appropriate.

(3) Drain

The main purpose of drainage ditches is to divert water, which is the water flow of various sources within the range of the subgrade (such as side ditches, cut ditches, pits, slopes and water near the subgrade), to designated locations outside the range of the bridge and culvert or subgrade. When the route is affected by multisection ditches or waterways, to protect the roadbed from water damage, drainage ditches can be set up or diverted to Xiaoqi to regulate water flow and waterways. The cross section of a drainage ditch is generally trapezoidal, and the size should be determined through hydraulic and hydrological calculations. The cross section size is determined according to the design flow rate. The bottom width and depth should not less than 0.5 m. The slope of the upper ditch is 1:1–1:1.5. The location of the ditch can be determined according to the need and combined with the local terrain and other conditions, as far as possible from the subgrade. The distance from the base of the subgrade slope should not less than 2 m, the alignment should be straight, the slope should also be as round as possible and made into an arc, the radius should at least 10–20 m, and the continuous length should be short, generally not more than 500 m.

The discharge of water from a gutter into another ditch or channel should be designed so that the original channel does not create scour or siltation. Generally, the ditch and the original channel should intersect at an acute angle; that is, the angle of intersection should be no more than 45°. If possible, a circular curve with radius R = 10b (b is the width of the trench top) can be used to connect the ditch with other channels downstream, as shown in Fig. 5.6.

The drainage ditch should have an appropriate longitudinal slope to ensure smooth flow, neither too steep, causing scour, nor too gentle, causing siltation. So it should be selected based on hydrologic-hydraulic calculations. Under normal circumstances, a slope of 0.5–1.0% is recommended, not less than 0.3% and not more than 3%. If the longitudinal slope exceeds 3%, corresponding reinforcement measures should be

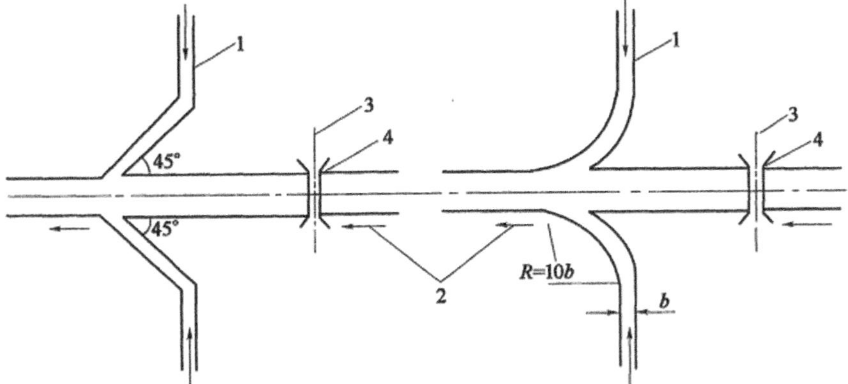

Fig. 5.6 Schematic diagram of drainage ditch and water channel connections

taken.

(4) Drops and rapids through

The drop and jet channel is a special form of subgrade surface drainage ditch, which is used for steep slopes with longitudinal slopes exceeding 10% and head height differences greater than 1.0 m. Owing to the steep longitudinal slope, fast water flow and large erosion force, the structure of the drop and rapid groove must be stable and durable. Typically, slurry blocks or precast cement concrete blocks should be used for building, and corresponding protection and reinforcement measures should be taken.

The structure of the falling water can be divided into single and multiple stages, and the bottom of the ditch can be divided into equal width and variable width. Single-stage flow is suitable for drainage ditch joints, where large water level drops, energy dissipation or changes in the direction of flow are required (Fig. 5.7) to address water excretion through culverts. Single-stage flow (equivalent to the rain well) samples in long steep slope zones of ditches are used to slow flow velocity and energy dissipation. Multiple-stage flow can be used as necessary, Fig. 5.8 shows an example. The width and length of each stage of the bottom can be the same symmetrical shape and can also be made according to the field conditions, width or unequal length and degree.

According to the characteristics of hydraulic calculations, the basic structure of a drop structure can be divided into three parts, namely, the inlet, damping pool and outlet, as shown in Figs. 5.10, 5.11, 5.12, and 5.13. The dimensions of each component are determined via hydraulic calculations. In general, if the geological conditions are good and the water table is low, the designed flow is less than 1.0–2.0 m³/s, and the height of the drop step (parapet) is not more than 2.0 m. Typically, the height of a platform is 0.4–0.5 m, the wall is built of a stone or concrete, the wall base burial depth is 1.0–1.2 times the water depth, but not less than 1.0 m, and should be deep below the frost line; the thickness of the stone wall is approximately 0.25–0.30 m. The damping pool plays a role in energy dissipation, which requires

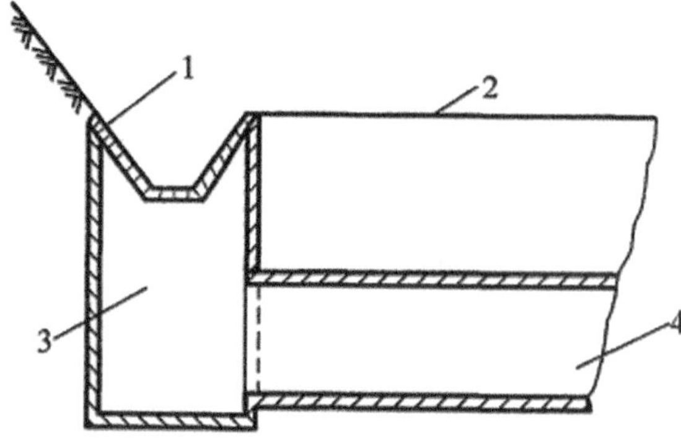

Fig. 5.7 Connection diagram of side ditch and culvert single-stage drop

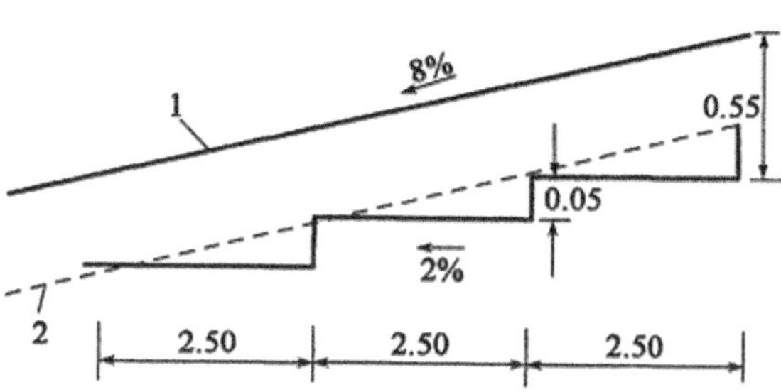

Fig. 5.8 Longitudinal profile of multistage drop

firm and stable energy. The bottom has a 1% longitudinal slope, the bottom thickness is approximately 0.30–0.35 m, the wall height should be at least 0.20 m greater than the calculated water depth, and the wall thickness should be similar to that of the parapet wall. The end of the pool is equipped with a damping sill, whose height depends on the calculation. The requirements are lower than the depth of the pool for the height of the wall 1/5–1/4 = (0.2–0.25)p, generally take c = 15–20 cm. The top thickness of the sill is approximately 0.3–0.4 m, and the bottom is reserved with a drainage hole of 5–10 cm to discharge the water in the tank when the water flow is interrupted.

The soil ditches at both ends of the drop structure should be strengthened to ensure that the water flows smoothly and prevent the water from scouring or silting to increase the drainage efficiency of the structure.

The mountain highway side ditch is a subgrade drainage channel and communicates up and down the back curve line outlet of a common drainage facility. The average longitudinal slope is steeper than existing, the strong stability of the structure is more demanding, the rapidity of the longitudinal slope depends on the terrain in the body slot, which generally can reach 67% (1:1.5). If the geological condition is good, when the need can also be steeper, but the structural requirements are stricter, the cost is also increased accordingly, and the design should be determined by comparison.

The jet stream trough is constructed of masonry (plaster) and cement concrete and can also be dug using rock slopes. In the case of immediate need, yes, nearby materials, bamboo and wood structure.

Figure 5.9 shows the structure of the jet stream trough. The hydraulic calculation is composed of three parts: the inlet, main groove (trough body) and outlet.

At the junction between the inlet and outlet of the jet stream groove and the main groove, the cross section of the groove is different. For smooth connection, a transition section can be set up, and the outlet section is provided with a damping pool. The dimensions of each part are determined via hydraulic calculation. When the design flow rate is not more than 1.0 m/s, the bottom of the pool tilt is a 1:1–1:1.5 small structure. The foundation of the jet stream groove must be stable, and the end and groove body should be equipped with ear walls at the bottom of the groove and

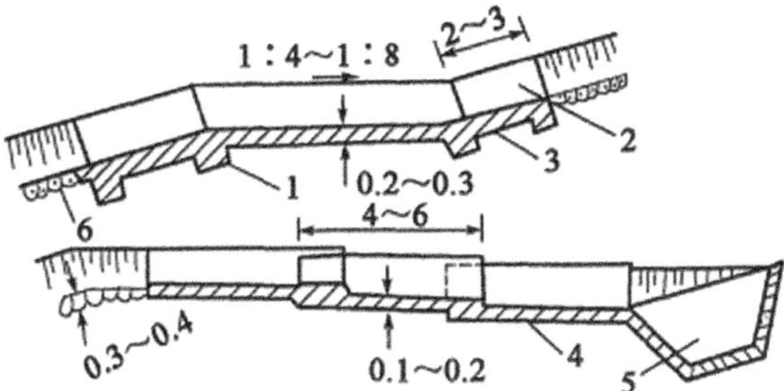

Fig. 5.9 Schematic diagram of the jet stream trough structure

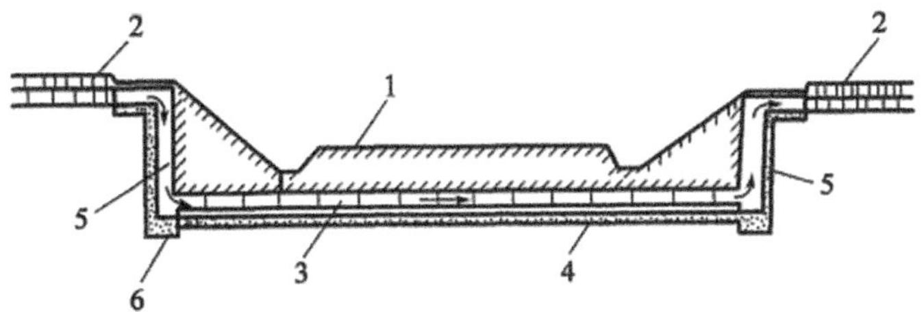

Fig. 5.10 Vertical shaft inverted siphon arrangement

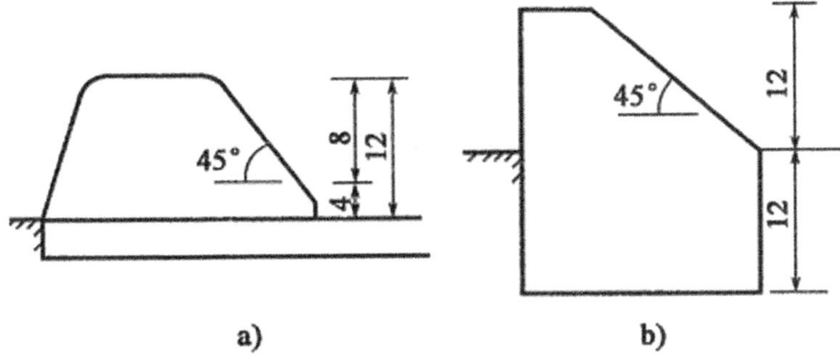

Fig. 5.11 Reference size of the cross section of the catchment belt

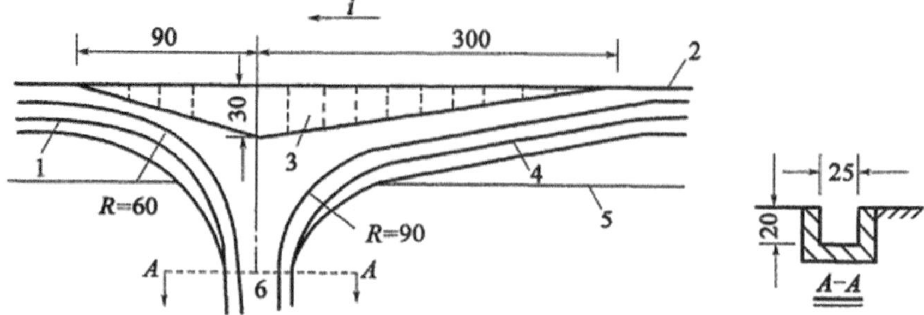

Fig. 5.12 Layout of the asymmetrical drainage outlet on the longitudinal slope section

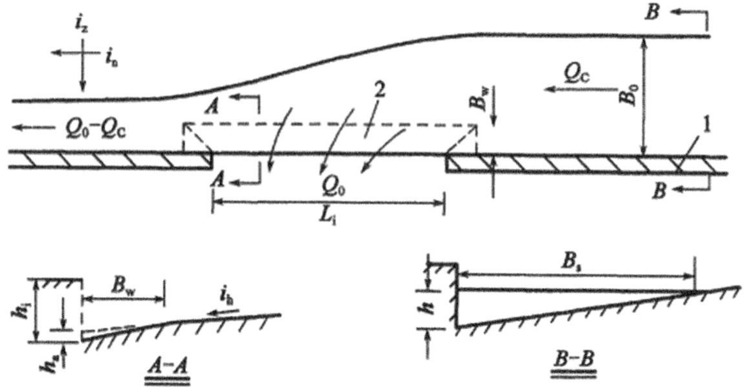

Fig. 5.13 Flow conditions around an open drain

buried under the ground every 2–5 m. When the groove body is long, it should be built in sections, each section should be approximately 5–10 m long, expansion joints should be preserved, and joints should be filled with waterproof materials.

(5) Inverted siphon and flume

When water needs to cross the roadbed, which is limited to the design elevation at the same time, it can use a pipe or groove, from the bottom of the subgrade or upper overhead crossing. As a pressurized drainage structure crossing beneath canals or aqueducts, the inverted siphon differs fundamentally from culverts in hydraulic behavior. It serves as critical infrastructure integrating irrigation and water conservancy systems [5].

The setting of an inverted siphon is usually because the subgrade crosses the original ditch and the water level of the target ditch is higher than the designed elevation of the subgrade, so the culvert cannot be set up under normal conditions. In this case, an inverted siphon is one of the feasible schemes, as shown in Fig. 5.10.

An inverted siphon refers to the use of potential energy to force the water to flow down through the water level difference between the upper and lower ditches, flow to the other side of the subgrade through the lower pipe of the subgrade, and then rise into the downstream channel. To design a pressurized pipeline system utilizing multiple vertical shaft inverted siphons, the system must account for complex flow direction changes and suboptimal hydraulic conditions, necessitating robust structural integrity. The pipeline should be designed to facilitate efficient sediment removal and maintenance while minimizing the number of vertical shafts. When vertical shafts are employed, rational design principles and rigorous hydraulic calculations should be applied to optimize the design scheme. Construction must adhere to stringent quality standards, with regular inspections and maintenance schedules implemented to ensure long-term operational reliability. There are two kinds of inverted siphon pipes: box shapes and round shapes. The main structures are cement concrete and reinforced concrete. Masonry structures can be used for temporary simple pipes, and steel pipes can be used for permanent or urgent needs. The pipe diameter is approximately 0.5–1.5 m, and the thickness of the filling soil on the roadbed near the pipeline is generally not less than 1.0 m to avoid the concentration of the driving load and pressure, and the cold area can also be anti-freezing. Considering the limited drainage capacity of inverted siphons and for the convenience of construction and maintenance, the pipeline should not be buried too deep, and the filling height should not exceed 3.0 m.

Vertical wells are set at both ends of the inverted siphon pipe, and the bottom elevation of the well is lower than that of the pipe to settle sediment and debris. It can also be changed to an inclined pipe or gentle slope to replace the vertical shaft lifting pipe. At this time, the water flow condition is improved, but the width of the subgrade increases, and the length of the pipe increases. To reduce the blockage phenomenon, the speed of the water flow in the pipeline should not be less than 1.5 m/s, and the sand sink and mud barrier should be set at the entrance.

(6) Evaporation pool

Arid climates, which are difficult drainage areas, can be used along centralized extraction pits or specially set up evaporation pools to remove surface water. A drainage ditch should be connected between the evaporation pool and the subgrade side ditch (or drainage ditch). The distance between the evaporation pool edge and the subgrade side ditch should not be less than 5 m, and the evaporation pool with a large area should not be less than 20 m^2. The water level in the pool should be lower than the bottom of the drain.

The capacity of the evaporation pool should be designed on the basis that the rainwater flowing into the pool from the subgrade sink can permeate and evaporate in time within a month. The water halo of each evaporation pool should not exceed 300 m^3, and the water storage depth should not exceed 2.0 m.

The evaporation pool should not be set up to cause salinization or swamping of the nearby ground.

Pavement Surface Drainage

The main task of pavement surface drainage is to quickly land on the road surface and road shoulder surface precipitation away so as not to cause water on the road surface and affect driving safety. The design of pavement surface drainage should be selected according to the following principles:

(1) Rain falling on the road surface should be discharged to both sides through the cross fall of the road surface to avoid water accumulation in the range of the pavement surface.

(2) When the longitudinal slope of the route is gentle, the catchment area is not large; the embankment is low; the side slope surface is not washed away; and transverse overflow flow should be adopted to remove surface water from the road surface.

(3) For high embankments with unprotected side slopes prone to water erosion or where existing slope protection remains vulnerable, an intercepting ditch shall be constructed along the shoulder edge. This structure collects roadway surface runoff and discharges it through slope drainage structures (chutes or pipes) away from the embankment.

(4) A catchment belt is set up to collect the surface water which with the water surface in the water section is arrested. On expressways and Class I highways, runoff should not overflow the outer edge of the right lane, and on Class II and lower-class highways, runoff should not overflow the centerline of the right lane.

Because the construction of catchment belts and jet troughs needs to increase project investment, it is necessary to analyze and compare the economy of investment. Analysis should use effective slope protection measures without catchment belts and jet troughs or to build catchment belts and jet troughs and reduce the cost of slope protection engineering requirements.

The catchment belt can be paved by asphalt concrete on site or by precast cement concrete blocks. When a precast cement concrete block with a catchment belt is used, the precast block should avoid affecting water drainage inside the road surface. The cross-sectional size of the catchment belt is shown in Fig. 5.11. The top surface of the catchment belt should be slightly higher than the design water surface height (water depth) h of the water cross section, and the design water depth should be calculated and determined according to the design flow formula (5.1).

$$Q_C = 0.377 \frac{1}{i_h n} h^{\frac{8}{3}} I^{\frac{1}{2}} \tag{5.1}$$

where Q_C is the drainage capacity of the ditch or pipe (m^3/s);

  i_h is the transverse slope of a trench or crossing section;

  n is the roughness coefficient of the ditch wall or pipe wall;

  h is the design depth (m);

  I is the hydraulic slope, which is the slope of the ditch or pipe to be used.

The outlet of the catchment belt can be arranged as an opening (bell mouth). To improve the drainage capacity, the outlet on the longitudinal slope section should be made into an asymmetrical flared outlet, and the outer side of the edge of the hard shoulder should be set up to gradually widen the low concave area. For the layout, see Fig. 5.12. The water discharge and length of the opening, width of the low concave area and depth of the concave area should be determined according to the hydraulic calculation of the drain.

For the open drainage outlet on the longitudinal slope section, the water discharge varies with the length of the opening L_i, the width of the lower concave zone B_W and the depth of the lower concave zone ha, and the longitudinal slope Iz and transverse slope of the water crossing section (Fig. 5.13).

Central Divider Drainage

The drainage of a central divider is an important part of the surface drainage of expressways and Class I highways. Different drainage methods should be selected according to the width of the divider, the form of green and traffic safety facilities, and the treatment of the surface of the divider. The "Code for Design of Highway Drainage" (JTG/T D33) has drainage requirements for the central separation belt as follows [6].

The width is less than 3 m, and the surface is drained by the central divider enclosed by the pavement. The surface water falling on the divider is drained to the carriageway on both sides, and its slope is the same as the transverse slope of the road surface. On superelevated section, a curb or drain can be set at the upper edge of the divider, or a straight gap circular water collection pipe or dish-shaped concrete shallow ditch and drain can be set in the divider (Fig. 5.14) to intercept and discharge surface water on the upper half of the road surface. The outlet of the water section of the edge stone can be an open-door type, grid type or combined type; the outlet of the shallow channel on the disc-shaped coagulation is a grille type. The grid bars should be parallel to the direction of water flow, the net drainage area of the orifice should account for more than half of the area of the lattice shed, and the spacing of the drain and the calculation of the intercepted discharge and the size of the section can be selected via calculation.

The width is more than 3 m, and the surface is not drained by the central divider with pavement-enclosed central divider. The surface water falling on the divider is collected in the low-lying place in the middle of the divider and discharged into the drainage outlet or the bridge and culvert channel across the road boundary through the longitudinal slope. The transverse slope of the divider should not be steeper than 1:6. The longitudinal drainage slope of the divider should not be less than 0.25% in the water-crossing section without pavement and 0.12% in the case of pavement. When the water flow velocity exceeds the maximum allowable velocity of the ground soil, the ground soil should be protected against scour within the width of the water cross section to make a triangular or U-shaped section of the ditch. The anti-scour layer can be lime or cement stabilizer or mortar flagstone paving, with a layer thickness of

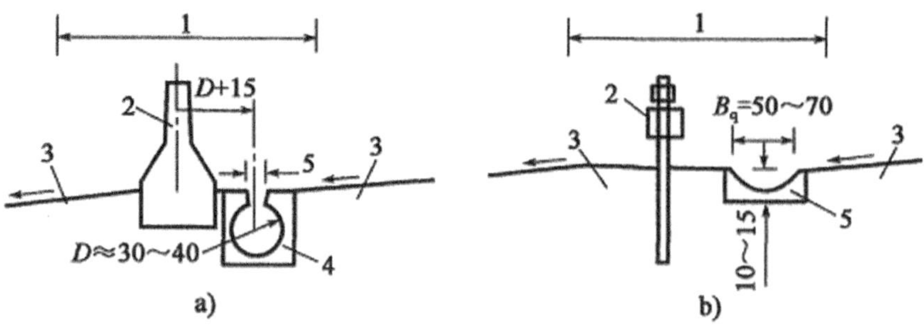

Fig. 5.14 Slot-type circular collector pipes or dish-shaped channels installed on superelevated sections

10–15 cm. When the water flow in the central divider is too high or if the flow rate exceeds the allowable range or when the water flow is collected in the low concave area of the divider, a grating inlet or drain should be set up and discharged to the bridge and culvert or the road boundary through the drain pipe. The grating may flush with the surrounding ground, or it may be lowered to create a depressed area within a certain width around it to increase the drainage capacity. In the longitudinal slope section, the discharge volume of the outlet can be calculated according to Eq. (5.1). For the grated drain at the bottom of the sag vertical curve, the water discharge is calculated according to Eqs. (5.2) and (5.3):

① When the water depth hi above the grating is less than 0.12 m:

$$Q_0 = 1.66 P_g h_i^{1.5} \tag{5.2}$$

where P_g is the effective perimeter length of the grating, which is half of the sum of the inlet perimeter length of the grating (m).

② When the water depth h_i above the grating is more than 0.43 m:

$$Q_0 = 2.96 A_i h_i^{1.5} \tag{5.3}$$

where A_i is half of the net drainage area of the grating inlet (m²).

When the water depth on the grating is between 0.12 and 0.43 m, the water discharge is between the results calculated via Eqs. (5.2) and (5.3) and can be obtained via linear interpolation according to the water depth.

For the surface of the central divider without pavement and without surface drainage measures, the surface water falling on the divider infiltrates, and underground drainage facilities are provided in the divider for drainage. The commonly used longitudinal drainage ditch should be separated by a certain distance through the transverse drainage pipe to drain the water in the sewer outlet. The seepage ditch is wrapped around the filter fabric (on the fabric) to prevent infiltration of fine particles carried by the water from blocking the seepage ditch. A geotextile impermeable layer coated with a double-layer asphalt is laid on the interface between the filler material and the pavement structure on the seepage trench. The drain pipe can be made of plastic pipe with a diameter of 70–150 mm.

5.3 Subgrade Drainage Design

5.3.1 Drainage Beneath the Subgrade

The water retained in the upper soil layer or shallow groundwater buried at a very small depth is called groundwater. Groundwater affects the strength of the subgrade or slope stability and should be set up underground drains (pipes), seepage trenches, inspection wells and other underground drainage facilities.

The commonly used subsurface drainage favilities for the subgrade are blind ditches, seepage trenches, seepage tunnels and seepage wells. The discharge capacity is not large, mainly in the seepage way to collect water flow, and the nearest discharge is outside the subgrade range. For a large flow of groundwater, a special underground pipeline should be set up to be excluded.

Because the subsurface drainage facilities are buried below the ground, they are not easy to maintain, and it is also difficult to determine the failure situation after the roadbed is built; thus, subsurface drainage facilities must be durable and effective.

(1) Blind ditches

Unlike the surface drainage of open ditches, blind ditches, also known as French drains, constitute a concealed work. From the structural characteristics of the blind ditch, owing to the ditch layer being filled with different sizes of granular materials, the use of seepage material permeates groundwater collection in the trench, and along the trench drainage to the designated location, this structure, relative to pipeline water, customarily called the blind ditch, has hydraulic characteristics belonging to turbulent flow.

Figure 5.15 shows a blind ditch set under a side ditch used to intercept the interlayer water flowing to the subgrade to prevent the slope of the subgrade from sliding and capillary rise, endangering the strength and stability of the subgrade.

Figure 5.16 shows a blind ditch installed beneath the side ditches on both sides of the road, which is used to lower the groundwater level and prevent the capillary rise into the working zone of the subgrade, resulting in moisture accumulation and causing

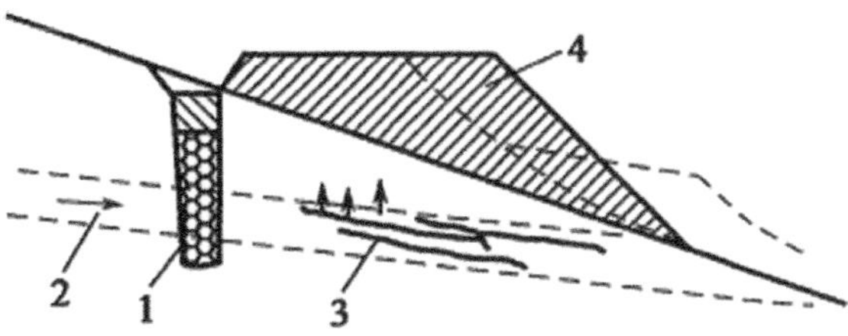

Fig. 5.15 A blind ditch installed beneath the side ditch

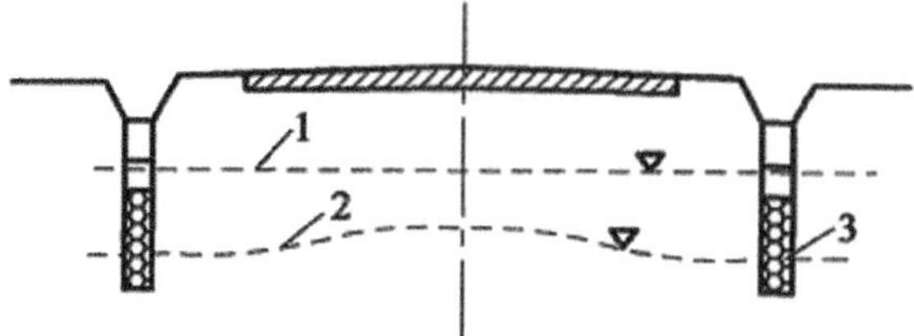

Fig. 5.16 A blind ditch installed beneath the side ditches on both sides of the road

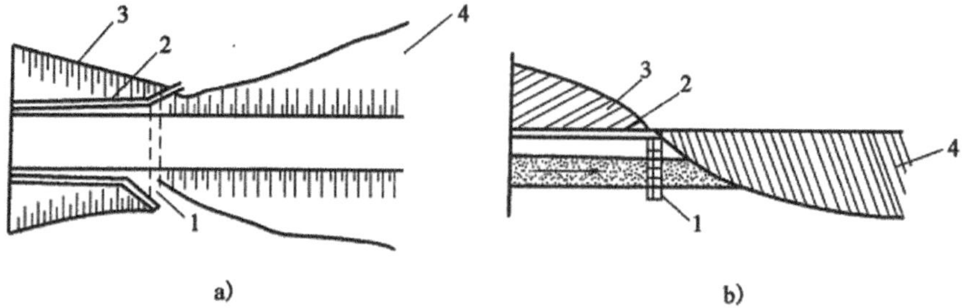

Fig. 5.17 Transverse blind ditch at the junction of cut and fill sections

frost heaving and mud pumping or reducing the strength of the upper subgrade due to excessive moisture.

Figure 5.17 shows a horizontal blind ditch located at the junction of cut and fill sections, which is used to intercept and discharge seepage water or minor spring water under the cutting and keep the embankment fill from water damage.

The blind trenches described above are all filled with granular material, which can be understood as simple blind ditches. Its structure is relatively simple, the cross section is rectangular, and it can also be made into a trapezoid with different top and bottom widths. The slope of the trench wall is approximately 1:0.2, the depth is approximately 1.0–1.5 m and the bottom width is approximately 0.3–0.5 m. From the middle of the bottom of the ditch, which is filled with coarse (3–5 cm) gravel, the voids are large, and water can flow through them. Coarse gravel on both sides and the upper side, according to a certain proportion of stratification (layer thickness of approximately 10 cm) filled with fine particle sizes of granular material, with a layer-by-layer particle size ratio of approximately 6 times the decrease in the top and bottom of the blind ditch, is generally provided with a thickness of more than a 30 cm impermeable layer, or the top is provided with a double layer of sod.

The drainage capacity of simple blind ditches is small and should not be too long. The bottom of the ditch has a longitudinal slope of 1–2%, and the elevation of the bottom of the outlet should be 20 cm higher than the highest water level outside the ditch to prevent water flow from seepage.

Blind ditches in cold areas should be treated with anti-freezing insulation or set below the freezing depth.

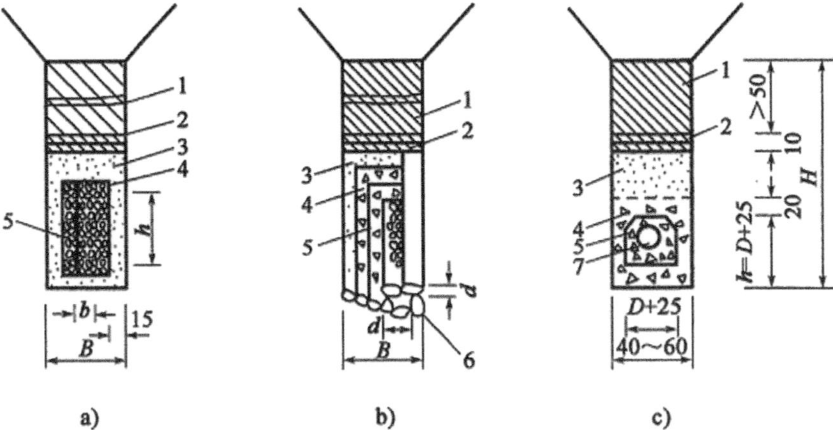

Fig. 5.18 Seepage ditch structures

(2) Seepage ditch

The groundwater is collected in the trench by means of infiltration, and the water is discharged to the designated location through the channel at the bottom of the trench. This kind of underground drainage equipment is collectively known as a seepage ditch. Its function is to lower the groundwater level or intercept groundwater, and its hydraulic characteristics are turbulent but different from those of simple blind ditches.

There are three structural forms of seepage ditches, as shown in Fig. 5.18.

The blind-ditch seepage trench is similar to the simple blind trench, but the structure is more perfect. When the groundwater discharge is large and the requirement is buried deeper, the hole or pipe can be set at the bottom of the trench. The former is called the seepage gallery, and the latter is called the seepage pipe trench.

(3) Seepage wells

Seepage wells are vertical underground drainage equipment under the local existence of a multilayer aquifer, which affects the roadbed with a thin upper aquifer. Displacement and horizontal seepage ditches are hard to arrange, can use vertical drainage (vertical), and can establish seepage wells through the impermeable layer and the roadbed upper groundwater within the scope of introducing deeper aquifers to lower the upper water table or eliminate it altogether.

The layout of the seepage well, as well as the diameter and seepage capacity, are determined via hydraulic calculations. Generally, the diameter of the seepage well is a 1.0–1.5 m cylindrical shape, and the side length is a 1.0–1.5 m square shape. The depth of a well depends on the geological formation. From the center to the periphery, the well is filled with coarse to fine sand or stone material, respectively, according to the level. The coarse material is permeable, and the fine material is filtered. The filling material was screened and washed. When it is applied, it should be separated by an iron sleeve and filled with materials of different particle sizes, the layers should be clear, and the coarse and fine materials should not be mixed to ensure that the seepage well can achieve the expected drainage effect.

Given that seepage wells are difficult to construct, the cost of the unit seepage area is greater than that of seepage ditches, which are generally less expensive. Sometimes, due to the high moisture content of the soil base, which severely affects the strength of the roadbed and road surface, other underground drainage equipment is not easy to arrange, and other technical measures, such as isolation layers, have a high cost. At this time, seepage wells can be used as one of the technical measures for design comparison and selection.

5.4 Integrated Drainage System Design

5.4.1 Separator Layer

Introduction

The separator layer is located between the base course and the subgrade (see Fig. 5.19). It can improve the moisture and temperature of the subgrade, protect the surface course and the base course from the adverse effects of changes in the moisture and temperature of the subgrade, and maintain the subgrade in a stable state [7–9]. At the same time, it can also distribute the load stress transmitted by the base, reduce the stress and deformation of the soil foundation, and prevent the subgrade soil from squeezing into the base.

According to their functions, the separator layers can be divided into drainage layer, isolation layer and anti-frost layer. The functions are as follows:

(1) Keep the subgrade soils from contaminating the permeable base and causing localized failure.
(2) Provide a stable working platform for construction.

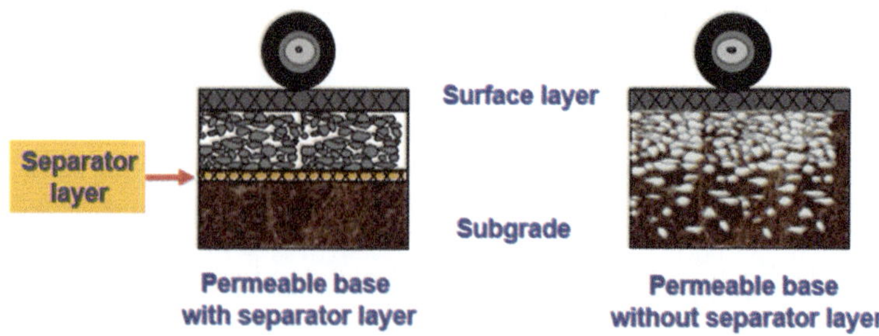

Fig. 5.19 Separator layer

Fig. 5.20 Improving drainage performance of the separator layer

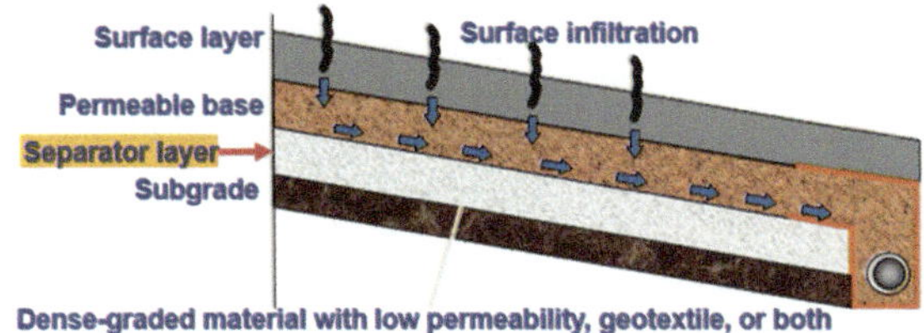

(3) Deflect infiltrated water into the edge drain in unique circumstances, allowing water to drain into the subgrade.

The separator layer should be provided for sections with subgrades under the following conditions: sections with high groundwater levels, poor drainage and subgrades, often in the wet state; soil cut sections with poor drainage; and rock excavation sections with poor moisture and temperatures conditions, such as fissure water and spring water. Seasonally frozen areas may produce frost heaving in the middle and wet sections, and the base course may be polluted. For example, the separator layer improves the drainage performance of the permeable base, as shown in Fig. 5.20.

Materials of the Separator Layer

The materials and thickness of the separator layer must be adequate for construction support and separation. For example, the geotextiles provide no construction support as a separator layer (Fig. 5.21).

The materials used as separator layers should be selected, and examples are listed as follows:

(1) Dense-graded aggregate.
(2) Dense-graded asphalt concrete.
(3) Cement-treated base (not recommended).
(4) Geotextiles.
(5) Asphalt chip seals (not recommended).
(6) Treated subgrade (not recommended).

Fig. 5.21 Geotextiles provide no construction support

Fig. 5.22 Aggregate/
geotextile combinations

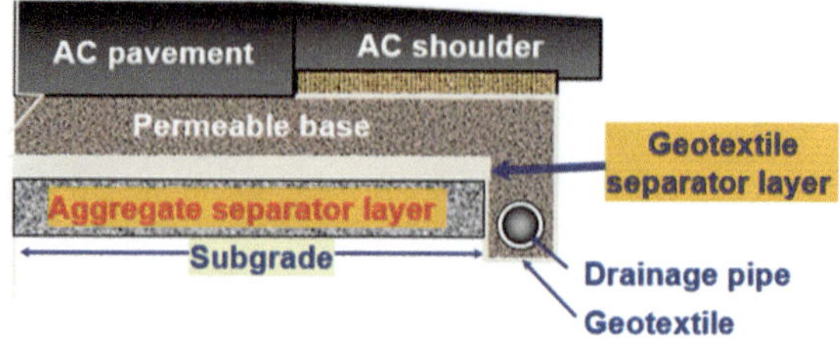

Several materials of the separator layer can be combined to ensure construction support and separation, as shown in Fig. 5.22.

Design and Construction of the Aggregate Separator Layer

These design rules shall be strictly followed when a dense-graded aggregate base material is used, and they are listed below:

(1) Gradation should prevent subgrade fines from being pumped into the permeable base.
(2) The percentage passing through a 0.075-mm sieve should be less than 12%.
(3) $C_U > 20$, preferably in the vicinity of 40.
(4) D_{15} (separator layer) $< 5*D_{85}$ (subgrade).
(5) D_{50} (separator layer) $< 25*D_{50}$ (subgrade).
(6) Class C aggregate AASHTO M 283.
(7) L. A. abrasion not to exceed 50%.

This separator layer should be constructed like any other granular subbase during the construction of the aggregate separator layer, and some construction experiences are listed below:

(1) Must be compacted in lifts not exceeding 100 mm.
(2) Should be compacted to 95% of the maximum dry density (MDD).
(3) Should not experience any rutting or movement during construction.
(4) The minimum thickness for separation is 100 mm.
(5) The thickness can reach 300 mm if a weak subgrade is constructed.

As the geotextile serves as a special aggregate separator layer, several notes are listed below:

(1) Prolonged exposure to sunlight results in degradation.
(2) Place in a manner that prevents wrinkles, ripping, or tearing.
(3) Light equipment can be allowed on the geotextile, but traffic must be limited.
(4) Heavy equipment should not be allowed on the geotextile without adequate material cover.
(5) Place the permeable base material using a paver/spreader or by back dumping from a truck.

Fig. 5.23 Edge drains for
new pavements

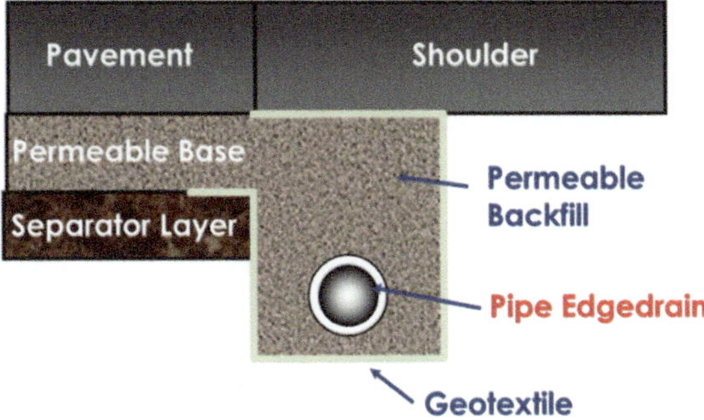

5.4.2 Edge Drains

Introduction

Edge drains are longitudinal drainage facilities positioned beyond the shoulders of
cut sections or downslope of low embankments. They function to collect pavement
surface runoff, drain intercepted hillslope water, convey flows into primary drainage
systems, and ultimately discharge through bridges/culverts [10–13]. It normally
contains various pipes and is used in both new and rehabilitation projects (Fig. 5.23).

Prefabricated Geocomposite Edge Drain (PGED)

Prefabricated geocomposite edge drains (PGEDs) are commonly used in rehabilitation projects, and the textures of materials and patterns of use are shown in Figs. 5.24
and 5.25.

Maintenance of Drainage Systems

The typical maintenance techniques are as follows:

(1) Daylighted bases.
(2) Keep the daylighted portion free of debris and vegetation.
(3) Flush with moderate-pressure water to remove silt deposits along the edge of
      the permeable base.

   When facing a project that contains permeable bases with edge drains, please
follow the listed maintenance methods:

(1) Keep outlets free of vegetation, debris, and sediment.
(2) Flush drains with high-pressure water.
(3) Replace damaged outlets, headwalls, and markers.

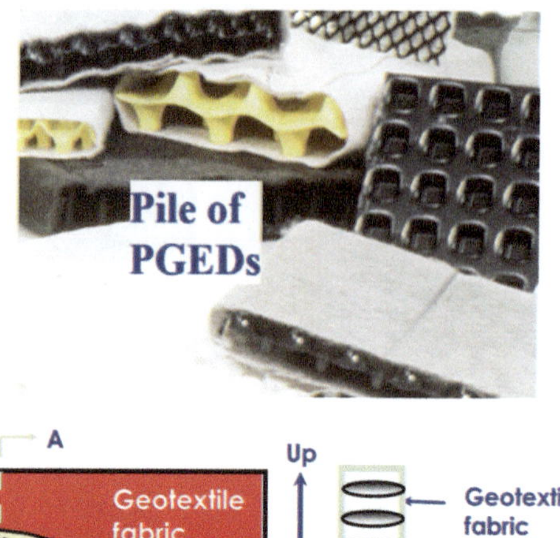

Fig. 5.24 Textures of the materials and patterns of PGED

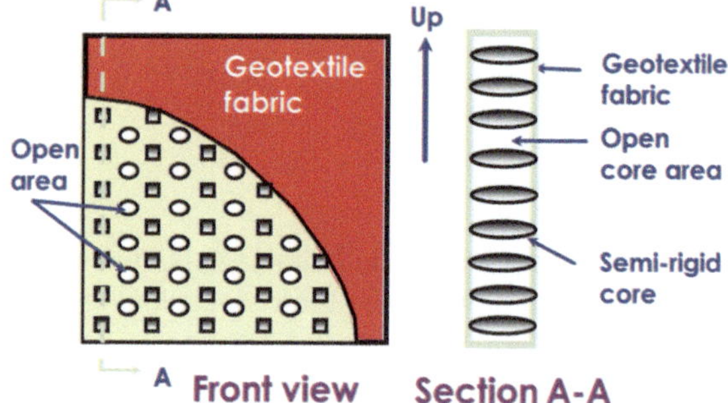

At present, performance monitoring tests exist for the water flowing in a drainage system (Fig. 5.26).

(1) Laboratory tests

    1. Constant-head test.
    2. Falling-head test.

(2) Field tests

    1. Percolation test.
    2. The field permeability testing device is shown in Figs. 5.5, 5.6, 5.7, 5.8, 5.9, 5.10, 5.11, 5.12, 5.13, 5.14, 5.15, 5.16, 5.17, 5.18, 5.19, 5.20, 5.21, 5.22, 5.23, and 5.24.

(3) Visual observation

The outlets and ditches were observed immediately after rainfall, as shown in Fig. 5.27.

Fig. 5.25 Pipe locations within the edge drains of different roads

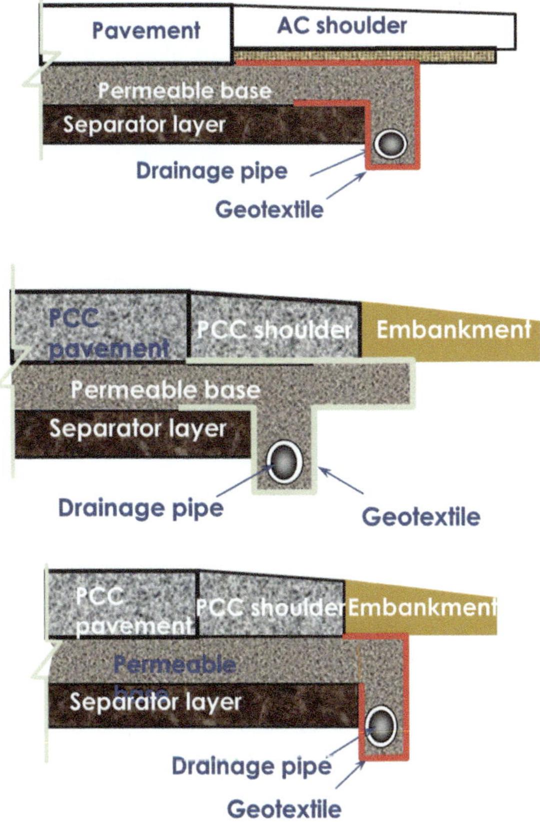

Fig. 5.26 Field tests of water flow in the drainage system

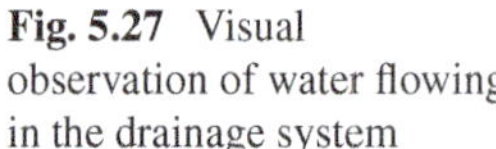

Fig. 5.27 Visual observation of water flowing in the drainage system

5.5 Simulation Method for Pavement and Subgrade Drainage Design

5.5.1 Introduction

"Stormwater management" refers to the cooperative efforts of public agencies and the private sector to mitigate, abate, or reverse the adverse effects of stormwater runoff, in terms of both water quantity and water quality [14–16]. Figure 5.28 shows the water-quality control through the first 3 mm of rainfall runoff diversion.

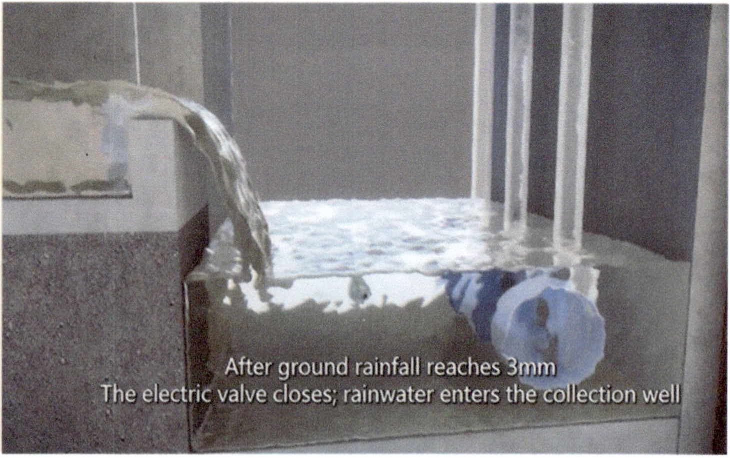

Fig. 5.28 Water-quality control through the first 3 mm of rainfall runoff diversion

5.5.2 *Highway: Fully Permeable Pavement Shoulder*

Fully permeable pavements capture stormwater and infiltrate it into the ground and/ or temporarily store it before discharging into the stormwater system, as shown in Fig. 5.29.

Fully permeable pavements offer benefits, including the following:

(1) Reducing the need for stormwater conveyance infrastructure.
(2) Capturing pollutants.
(3) Reducing air temperature.
(4) Being cost-competitiveness compared with other LID practices.

Fully permeable pavements capture stormwater and infiltrate it into the ground and/or temporarily store it before discharging it into stormwater conveyance system, as shown in Fig. 5.30.

A study on fully permeable pavements in California was performed. Three locations, namely Eureka, Sacramento and Riverside, in California were selected. Four parameters, namely, the subgrade soil hydraulic conductivity, recurrence interval, geometric parameters and initial water content, were considered, as shown in Fig. 5.30. Finally, several conclusions have been reached: (1) An aggregate base thichness of approximately 1.5 m was adequate for most two-lane highways in California. (2) The saturated hydraulic conductivity is the most important parameter.

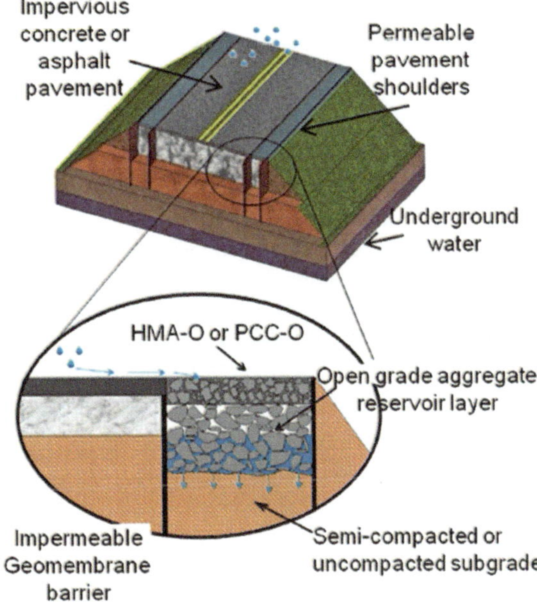

Fig. 5.29 Fully permeable pavement shoulder

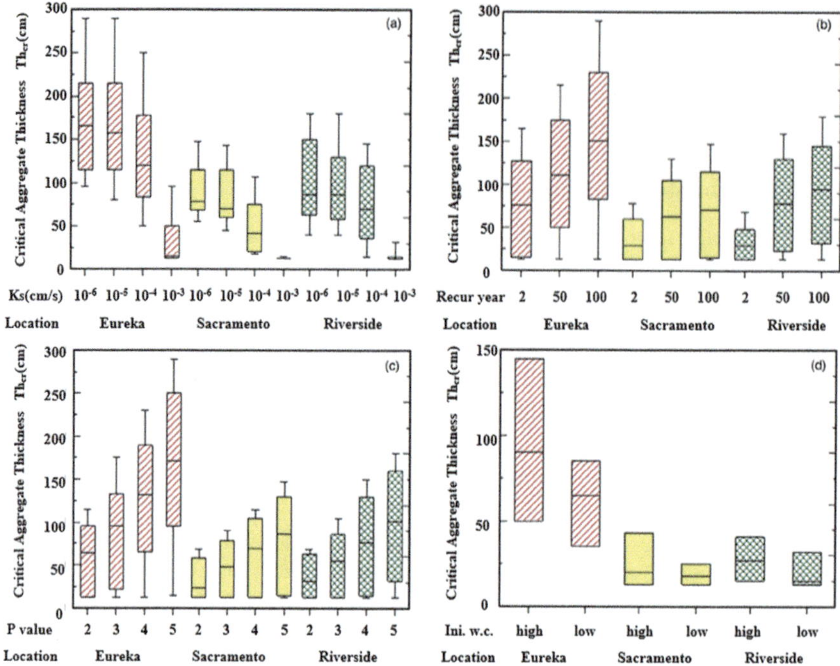

Fig. 5.30 Influence of parameters on the critical aggregate base thickness

5.5.3 *Urban Pavement and Subgrade*

Permeable pavements are usually adopted in sidewalks, plazas, parking lots, etc. [17–19]. For stakeholders, one of their main concerns is how much flooding and pollutants can be reduced after adopting permeable pavements (Fig. 5.31). The benefits can be predicted through simulation, as shown in Fig. 5.32. The Storm Water Management Model (SWMM) invented by the Environmental Protection Agency (EPA) is a dynamic rainfall–runoff simulation model used for the simulation of runoff quantity and quality. EPA SWMM has been one of the most widely used software programs around the world.

Fig. 5.31 Mechanism of flood and pollutant reduction by permeable pavements

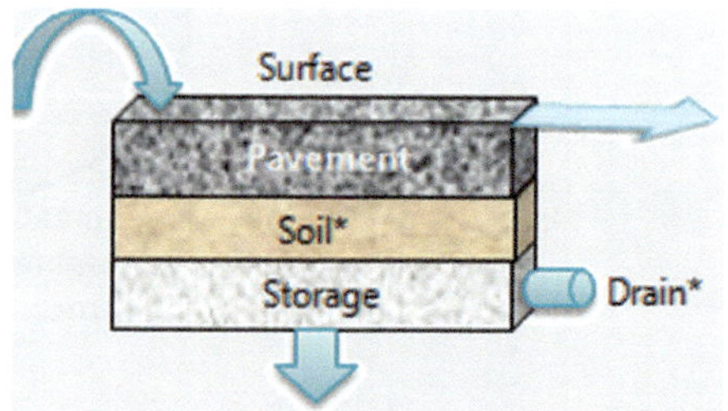

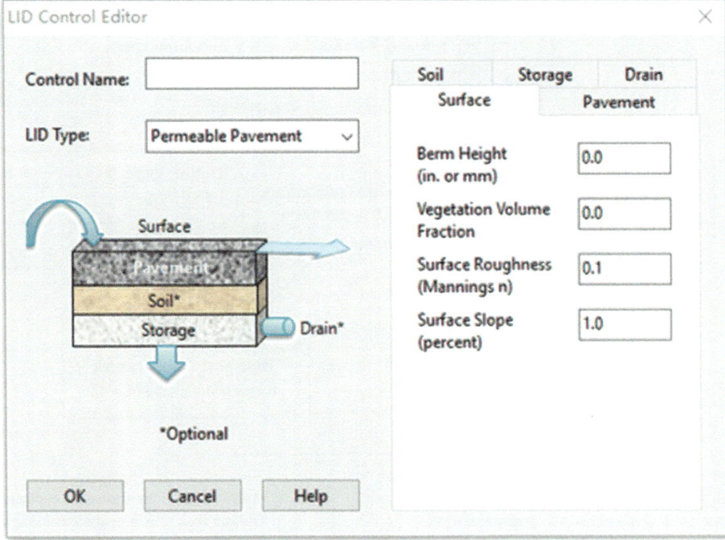

Fig. 5.32 Simulation-based evaluation of the benefits of permeable pavements

5.5.4   Case Study: Tongji Siping Campus Using SWMM

Some basic data still need to be obtained when SWMM is used to simulate the study area, as shown in Fig. 5.33. They are listed as follows:

(1)   Digital elevation map.
(2)   Land use plan.
(3)   Data of the underlying surface.
(4)   CAD drawings of Sewer system.

Fig. 5.33 Tongji siping campus

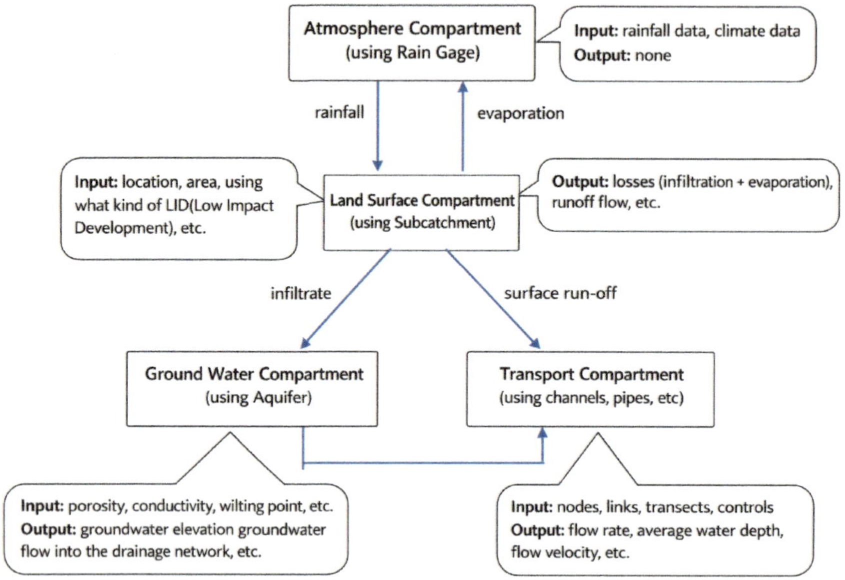

Fig. 5.34 SWMM compartment, inputs and outputs

Compartments, inputs and outputs are the key technologies of the SWMM, as shown in Fig. 5.34. The detailed information is as follows:

(1) Atmospheric Compartments: Rainfall data and climate data (including temperature, wind speed, and evaporation data).
(2) Land Surface Compartments: Location, area, width, low-impact development, etc.
(3) Groundwater Compartment: Water content, hydraulic conductivity, wilting point, etc.
(4) Transport Compartments: Nodes, links, transects, controls, etc.

A study on runoff quantity and water quality after rainfall events was performed on the urban roads of the Tongji Siping Campus, China, via SWMM software.

The runoff coefficient and peak runoff were selected and analyzed (Fig. 5.35). Several conclusions have been reached: (1) Adopting permeable pavements can effectively reduce runoff coefficients and peak runoff; (2) Compared with increasing the thickness of permeable pavements, increasing their area is more effective.

Note: (1) The basic scenario does not adopt permeable pavements; (2) Scenarios 1–5 differ in materials, structures and areas of permeable pavements.

TSS reduction and TN reduction were selected and analyzed, as shown in Fig. 5.36. Several conclusions have been reached: (1) The heavier the rainfall, the more pollutants will be washed off by runoff; (2) The pollutant reduction effect varies according to the materials, structure and area of permeable pavements.

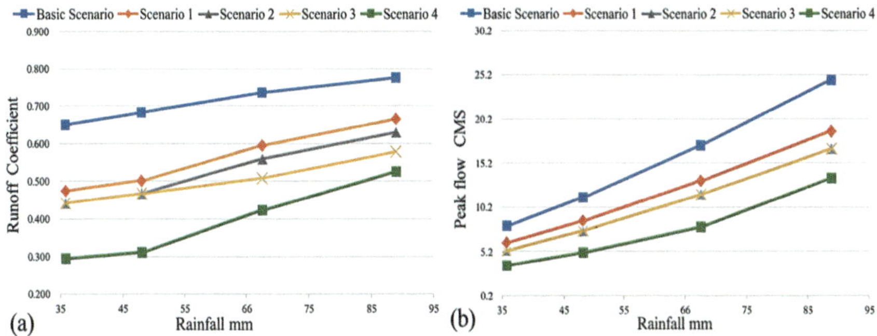

Fig. 5.35 Runoff coefficient and peak runoff

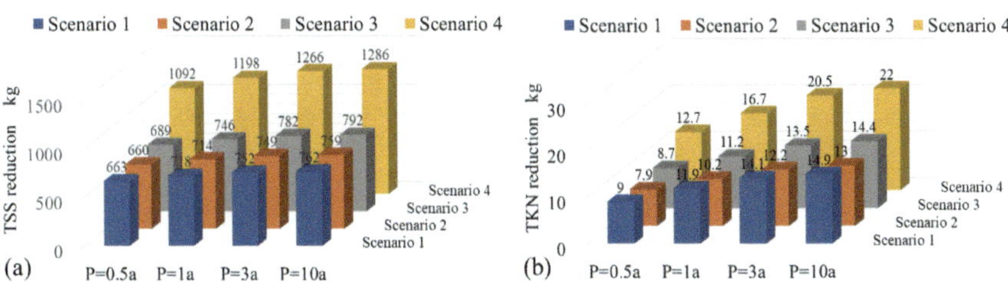

Fig. 5.36 TSS reduction and TN reduction

Note: (1) The basic scenario does not adopt permeable pavements; (2) Scenarios 1–5 differ in materials, structures and areas of permeable pavements.

Discussion, Questions and Exercises

(1) Describe the main sources of water affecting road subgrades and their impact on road stability. How does the drainage design address these effects?

(2) What are the main functional requirements of slope drainage facilities in road design? Discuss the design considerations for their layout, materials, and performance.

(3) How are central divider drainage systems designed for varying road conditions, and what factors influence the selection of appropriate methods?

(4) What is the role of blind ditches in underground drainage systems? Describe their structure, functionality, and advantages in specific geological conditions.

(5) Explain the purpose and functions of separator layers in road construction. What materials are commonly used, and how do they address specific subgrade challenges?

References

1. FENG B, LI Y. Analysis on the Design of Bridge Deck Drainage Control[J]. Inner Mongolia Science & Technology and Economy, 2015(08): 91–92
2. SHI W. Highway Roadbed Drainage Design Study[J]. Transportation World, 2022 (9): 108–109 + 132
3. Recommended Industry Standards of the People's Republic of China. JTG/T D33– 2012 Code for Design of Highway Drainage [S]. Beijing People's Communications Publishing House, 2013
4. LIU J P. Asphalt Highway Subgrade Pavement Drainage Design Study[J]. Transportation World, 2021 (34): 37–39
5. LI J Y, ZOU J R, WANG G, et al. Experimental Study on Drainage Performance of Subgrade Seep Ditch [J]. Journal of Hydraulic and Architectural Engineering, 222,20(02):222–228
6. ZHANG Y. Median Fill Roadbed Drainage Design Analysis [J]. Transportation world, 2022 (15): 155–156
7. DEBO T N, REESE A. Municipal Storm Water Management[M]. United States: The Chemical Rubber Company (CRC) Press, 2002
8. LINDLY J K, ASHRAF E. Open-graded Highway Bases Make Perm Eam Eter Setup Important[J]. Journal of Transportation Engineering, 1998, 124(2): 144–148
9. ROSSMAN L A. Storm Water Management Model User, Manual, Version5.0[M]. Cincinnati: National Risk Management Research Laboratory, 2010
10. AHIABLAME LM, ENGEL BA, CHAUBEY I. Effectiveness of Low Impact Development Practices: Literature Review and Suggestions for Future Research. Water, Air, and Soil Pollution, 2012, 223: 4253–4273
11. REVITT D M, SCHOLES L, ELLIS J B. A Pollutant Removal Prediction Tool for Stormwater Derived Diffuse Pollution[J]. Water Science and Technology, 2008, 57(8): 1257–1264
12. JENNIFER M, TERRY L. Practical Review of Pervious Pavement Designs[J]. Acta Hydrochimica Et Hydrobiologica, 2014, 42(2):111–124
13. STORMONT J C, ZHOU S X. Improving Pavement Subsurface Drainage Systems by Considering Unsaturated Water Flow[R]. Department of Civil Engineering, University of New Mexico, Aibuquerque, New Mexico,2001:1–7
14. TEMPRANO J, O ARANGO, CAGIAO J, et al. Stormwater Quality Calibration by SWMM: A Case Study in Northern Spain[J]. Water S A, 2006, 32(1):55–63
15. PAGOOTTO C, LEGRET M, LE CLOIREC P. Comparison of the Hydraulic Behavior and the Quality of Highway Runoff Water According to the Type of Pavement. Water Research. 2000, 34(18): 4446–4454
16. PALLA A, GNECCO I. Hydrologic Modeling of Low Impact Development Systems at the Urban Catchment Scale [J]. Journal of Hydrology, 2015, 528: 361–368
17. TSIHRINTZIS V A, HAMID R. Runoff Quality Prediction from Small Urban Catchments using SWMM[J]. Hydrological Processes, 2015, 12(2): 311–329
18. PARK J, YOO Y, PARK Y, et al. Analysis of Runoff Reduction with LID Adoption using the SWMM[J]. 2008, 24(6): 805–815
19. DO-HYSON P, YOUNG-HWAN L, JIN-KYU C, et al. Study on the Runoff Characteristics of Nonpoint Source Pollution in Municipal Area Using SWMM Model -A Case Study in Jeonju City[J]. Journal of Korean Society for Geospatial Information System, 2005, 14(12)

Chapter 6
Subgrade and Pavement Construction

6.1 Introduction

While ideal designs guide construction, subgrade engineering involves multiple flexible factors. Particularly, the complex internal structure of rock/soil makes perfect design impossible during planning stages. Thus, construction becomes the critical phase for refinement. In particular, the internal structure of rock and soil is complex and changeable, and it is difficult to perfect the design stage, which must be further improved in the construction process. "Meticulous design, careful construction" is a complete process. In terms of manpower, resources and financial resources, as well as speed, efficiency and safety requirements, construction is more important and complicated than design is. The amount of subgrade earthwork is large, and the distribution is uneven, which is restricted not only by facilities related to subgrade engineering, such as subgrade drainage, protection and reinforcement but also by other highway engineering projects, such as bridges, culverts, tunnels, pavements and auxiliary facilities. In addition, many subgrade projects, such as earthwork, stonework and masonry, have their own characteristics in terms of construction methods and technical operations. Therefore, the construction methods, quality standards, technical operations, and construction management of subgrade construction vary.

Pavement construction critically impacts pavement performance and service life. Good pavement structure combination design, material design and thickness design provide technical guarantees for the extension of pavement service life, and pavement construction is the last step in realizing these technologies. Pavement construction is a systematic project, and the final quality of pavement construction is related to most aspects of the construction process. Pavement construction must carry out a reasonable construction organization design, and the pavement design, management, supervision and construction units must coordinate and cooperate to achieve careful design, conscientious construction, and strict supervision. Pavement construction must be checked at every level, strict requirements must be met, the construction process must be optimized as much as possible, and the construction quality must

© Tongji University Press Co., Ltd. 2026

H. Li et al., *Road Engineering*, https://doi.org/10.1007/978-981-95-6659-4_6

be improved. In pavement construction, it is necessary to ensure that the quality of the raw materials is sufficient, the mixing ratio is accurate, the mixing ratio is uniform, the paving process is smooth, the rolling process is compact, and the joints are smooth to ensure the engineering quality of the pavement.

6.2 Subgrade Construction

6.2.1 Embankment Construction and Compaction

Basic Requirements

For the excavation and filling of the soil subgrade, workers must install drainage systems prior to excavation, including excavation of temporary drainage grooves on the ground and ways to lower the groundwater level, to keep the construction site dry. To effectively control soil moisture content, the construction work surface of the soil subgrade should not be too large to facilitate the organization of rapid construction, excavation and transportation; timely filling and compaction; and reducing the amount of sun and rain during the construction process. To maintain natural humidity, dry or wet conditions should be avoided. Under normal conditions, the natural moisture content of the soil is close to the optimum value. If necessary, artificial watering or drying measures should be considered. During the rainy season, temporary drainage should be strengthened in accordance with the relevant provisions of the technical operating procedures to ensure the quality of the subgrade. Springing effect: rebound deformation due to excessive moisture after rolling, so it must be excavated and refilled, and other corresponding reinforcement measures can be taken if necessary.

Surface obstacles within the subgrade excavation and filling range should be removed in advance, including the demolition of the original house, the removal of the stems and roots of trees and jungles, and the removal of the surface planting soil and the sundries stipulated in the design documents or regulations. Under this premise, the upper layer of the embankment should be reinforced according to the design requirements if necessary.

Under normal circumstances, the soil embankment should be filled in layers and fully compacted within the full width. At the end of daily construction, the surface fill should be compacted to prevent rain or exposure during intervals. The thickness of the layer depends on the compaction tool, and the general compaction thickness is 20–25 cm. Where the embankment is widened or where the old and new soil layers overlap, the original soil layer should be dug into steps and filled with new soil layer by layer.

Filling Scheme

Soil embankments (including stone soil) can be divided into two construction schemes according to the filling sequence: layered tiling and vertical filling. Layered tiling is the basic scheme. If it meets the requirements of layered filling and compaction, the effect is better, and the quality is guaranteed. It should be used as much as possible when conditions permit. Vertical filling is a scheme adopted for local embankments under certain conditions.

(1) Layered tiling

Layered tiling is conducive to compaction and can ensure that different soils are filled according to the specified layers. Figure 6.1 shows a schematic diagram of the filling scheme with different soils. Figure 6.1a shows the correct filling plan: different soils are layered horizontally to ensure uniform strength; soils with poor water permeability, such as cohesive soil, should generally be filled in the lower layer, and the surface should be a two-way transverse slope, which is beneficial for eliminating accumulated water and preventing water damage; when there are different soils in the same layer, the splices should be inclined to ensure that the strength is relatively uniform within the thickness of the layer and prevent obvious deformation. Figure 6.1b indicates an incorrect filling scheme: not layered horizontally, with water accumulated on the reverse slope, with frozen soil blocks and coarse rocks, steep slopes, etc., which may easily lead to uneven strength of the embankment and unfavorable drainage. In addition, embankment-filling materials should not contain harmful impurities (vegetation, organic matter, etc.) or untreated inferior soil (fine silt, expansive soil, saline soil, humus, etc.). Sandy soil is suitable for backfilling structures such as bridges and culvert retaining walls. It can prevent uneven settlement and pile up backfill and exhaust according to relevant regulations.

(2) Vertical filling

Vertical filling refers to progressively deep filling forward along the direction of the road centerline, as shown in Fig. 6.2. When the route crosses a deep valley or pond, the ground height difference is large, the filling area is small, and it is difficult to unload it horizontally in layers. The subgrade is filled and excavated on the steep slope section, and the cross slope of the partial road section is steep or difficult to fill in layers, etc. Filling plans. The quality of vertical filling lies in the compactness of the filling soil, and necessary technical measures are adopted for this purpose. If a vibrating or hammer tamper is used, sand and gravel fillers with less settlement and a more uniform particle size are used; the full width of the embankment is formed at one time; higher-level road surfaces are not temporarily built, and natural settlement is allowed in a short period of time. In addition, a mixed-filling scheme should be adopted as much as possible; that is, the lower layer should be filled vertically, and the upper layer should be horizontally layered. If necessary, measures such as injection, reaming or dynamic compaction can be considered with reference to foundation reinforcement to ensure that the filling has sufficient compactness.

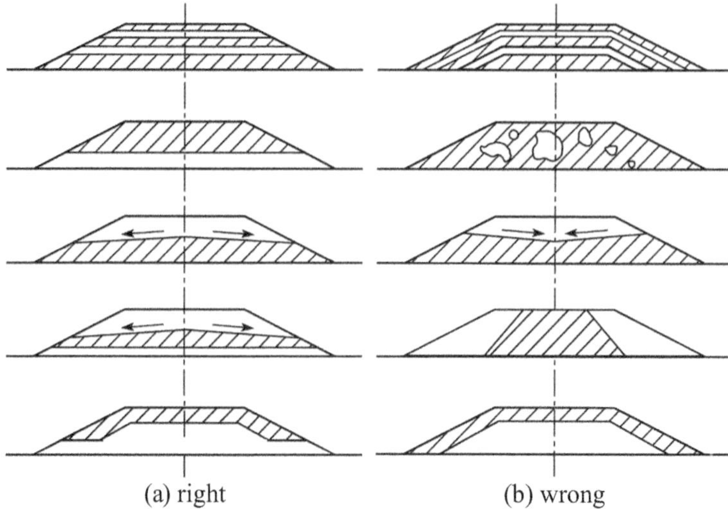

(a) right   (b) wrong

Fig. 6.1 Schematic diagram of the soil embankment filling scheme

Fig. 6.2 Schematic diagram
of the vertical filling scheme

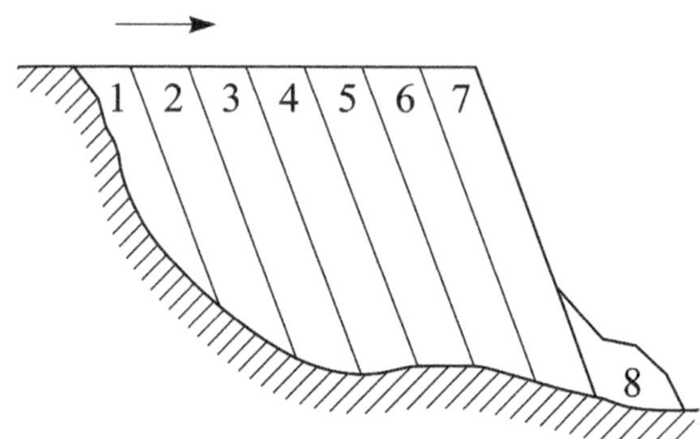

Subgrade Compaction

Subgrade construction destroys the natural state of the soil, causing its structure
to loosen and particles to recombine. For the subgrade to have sufficient strength
and stability, it must be compacted to improve its compactness. Therefore, subgrade
compaction is an important process in the subgrade construction process, and it is
also a fundamental technical measure to improve the strength and stability of the
subgrade [8].

(1) Selection and operation of compaction tools

The choice of compaction equipment and reasonable operation also affect the
subgrade compaction effect.

There are many types of subgrade compaction equipment, which can be roughly divided into three types: rolling type, ramming type and vibration type. The rolling type (also known as the static rolling type) includes smooth rolling (ordinary two-wheel and three-wheel rollers), Sheepfoot rolling and pneumatic tire rolling. In the ramming type, in addition to the artificial use of stone and wood, rammers, rammers, pneumatic rammers and frog rammers are used as motor equipment. There are vibrators, vibratory rollers, etc., of the vibratory type. In addition, automobiles, tractors and earth-moving machinery in earth-moving tools can also be used for subgrade compaction.

Different compaction tools are suitable for different soil qualities and thicknesses of soil layers, which is also the main basis for the selection of compaction tools. Table 6.1 lists the general technical performance of several commonly used compaction tools. Under normal circumstances, for the compaction effect of sandy soil, the vibration type is better, the ramming type is second, and the rolling type is poor; for clay soil, the rolling type or ramming type should be used, and the vibration type is poor. Not even valid. Different compaction tools are suitable for a certain compaction thickness and number of compactions under the condition of the optimum moisture content of the subgrade. Table 6.2 presents the suggestions for various types of compacting machines suitable for improving soil quality.

The external force exerted by the compactor on the soil should be controlled to prevent excessive compaction, resulting in failure and waste. It is generally believed that the unit pressure during compaction should not exceed the ultimate strength. The ultimate strength of different soils is related to factors such as the weight of the compactor, the mutual contact area, the loading speed and the action time (number of passes). Table 6.3 lists the strengths of several types of compaction tools when they act on different soils under the condition of optimum moisture content, which can be used for reference when tools are selected and compaction functions are controlled.

Practical experience has shown that when the subgrade is compacted, under the conditions that the type of equipment, the thickness of the soil layer and the number of rolling passes have been selected, the compaction operation should be light and then heavy, slow then fast, edge first and then middle (high road sections, etc.), then it should be low first and then high). When compacting, the two adjacent wheel tracks should overlap 1/3 of the wheel width to keep the compaction uniform and prevent leakage. During the whole process of compaction, the moisture content and compactness should be checked frequently to meet the requirements of the specified compaction.

(2) Subgrade Compaction Standard

In the field construction of subgrades, due to various conditions, the maximum dry weight (γ_0) obtained from the indoor standard compaction test should be appropriately reduced. The severity γ value measured on site should conform to the following formula:

$$\gamma = K \cdot \gamma_0 \tag{6.1}$$

Table 6.1 Technical performance of the road roller

Machine name	Maximum effective compaction thickness (solid thickness) (m)	Number of rolling trips			Applicable soil types
		Clay soil	Silty soil	Sandy soil	
Artificial ramming	0.10	3–4	1–3	2–3	Clay and sandy soils
Traction smooth grinding	0.15	–	7	5	Clay and sandy soils
Sheepfoot roller (2)	0.20	10	6	–	Clay soils
Automatic smooth grinding 5t	0.15	12	7	–	Clay and sandy soils
Automatic smooth grinding 10t	0.25	10	6	–	Clay and sandy soils
Pneumatic tyre roller 25t	0.45	5–6	3–4	1–3	Clay and sandy soils
Pneumatic tyre roller 50t	0.70	5–6	3–4	2–3	Clay and sandy soils
Tamping machine 0.5t	0.40	4	2	1	Sandy soils
Tamping machine 1.0t	0.60	5	3	2	Sandy soils
Ramming board 1.5t, drop high 2 m	0.65	6	2	1	Sandy soils
Caterpillar	0.25	6–8	6–8	6–8	Clay and sandy soils
Vibrating	0.40	–	2–3	2–3	Sandy soils

where

γ is the site-measured dry weight.

K is compactness.

The compaction degree K is the subgrade compaction standard stipulated in the current code. The correct selection of the K value is related to the stress state of the subgrade, the design requirements of the subgrade and pavement, and the construction conditions. It must consider needs and possibilities and pay attention to practical effects and the economy.

Figure 6.3 is a schematic diagram of the relationship between the stress σ in the soil and the depth Z when the subgrade is stressed, which shows that the subgrade surface is most affected by the driving force, and the force decreases sharply from the top down. Its influence depth is in the range of 1.0–2.0 m. When Z is larger, the subgrade mainly bears the gravity of the subgrade itself. Therefore, the compaction degree of the subgrade fill should be gradually improved from bottom to top.

The higher the road grade is, the higher the requirements for the strength of the subgrade; the worse the natural conditions are, the more unfavorable the strength and

Table 6.2 All kinds of suitable soil rolling machinery

Machine name	Soil classification				Note
	Fine grained soil	Sand soil	Gravel soil	Coarse-grained soil	
Two wheel light roller 6–8t	A	A	A	A	Used for prepressing and leveling
Two wheel light roller 12–18t	A	A	A	B	The most commonly used
Tyre roller 25–50t	A	A	A	A	The most commonly used
Sheepfoot roller	A	C/B	C	C	Silty, clay sand is available
Vibratory roller	B	A	A	A	The most commonly used
Bump vibratory roller	A	A	A	A	It is best to use fine grained soil with higher moisture content
Hand vibratory roller	B	A	A	C	Used in cramped locations
Vibrating plate ram	B	A	A	B/C	Used in cramped locations, Mechanical weight >800 kN can be used for macrogranular soil
Walking vibration ram	A	A	A	B	Used in cramped locations
Rammer/ Ram plate	A	A	A	A	Ramming has the greatest effect on depth
Bulldozer, backhoe	A	A	A	A	It is only used for leveling soil layer and preloading

Notes 1. Symbols in the table: A means applicable; B means available without suitable machinery; C means not applicable

2. The types of soil are divided according to the current "Highway Geotechnical Test Regulations" (JTG E40) [1]

3. For special soil and loess (CLY), expansive soil (CHE), saline soil, etc., the compaction machinery selection can be considered fine-grained soil

4. Self-propelled road rollers should be used for compaction of general embankments and cutting bases, and linear forward and backward operations should be used

5. Sheepfoot rollers (including bump rollers and strip rollers) should be used in conjunction with smooth rollers

Table 6.3 Ultimate strength of the soil at different compaction times

Soil type	Compressive strength (MPa)		
	Smooth surface grinding	Pneumatic grinding	Rammer plate (diameter 70–100 cm)
Low viscosity soil (sandy soil)	0.3–0.6	0.3–0.4	0.3–0.7
Medium viscosity soil (silty soil)	0.6–1.0	0.4–0.6	0.7–1.2
High viscosity soil (clayey soil)	1.0–1.5	0.6–0.8	1.2–2.0

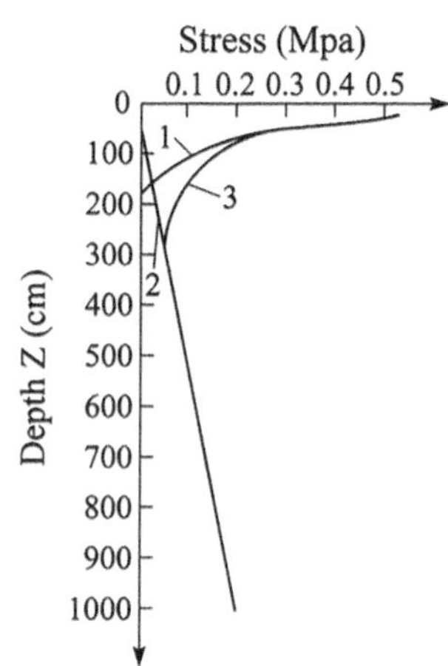

Fig. 6.3 Schematic diagram of the subgrade stress change curve with depth. 1—Driving load; 2—dead weight curve of the subgrade; 3—Superposition curves of the two

stability of the subgrade; the different excavations and fillings of the subgrade and the strength and stability of the subgrade are also different. On the basis of the above analysis, the subgrade compaction degree K stipulated in the "Code for Design of Highway Subgrade" (JTG D30) [5] is listed in Table 6.4.

The compaction degree listed in Table 6.4 is based on the heavy compaction test method of the Technical Standard for Highway Engineering (JTG B01) [2]. Since there is less rain in special arid areas and the groundwater level is also low, a slight reduction in the degree of compaction does not affect the firmness, stability or durability of the subgrade. It is indeed difficult for the subgrade to reach the optimum moisture content and compact to the specifications in Table 6.4. Therefore, the degree of compaction in special arid regions can be appropriately reduced. When asphalt concrete or cement concrete pavement is constructed on Grade 3 and Grade 4 roads, subgrade compaction should adhere to the Grade 2 road standard (Tables 6.5 and 6.6).

For rock-filled embankments, including embankments filled with blasted stones in layers, the compactness of the soil subgrade cannot be used to determine the compactness of the subgrade. China's Technical Specification for Highway Subgrade

Table 6.4 Subgrade compaction requirements

Subgrade area		Depth below the surface of the road (m)	Degree of subgrade compaction (%)		
			Highway, first class highway	The secondary roads	Third and fourth class highways
The road bed		0–0.3	$\geq$96	$\geq$95	$\geq$94
Under the road bed	Light, medium and heavy traffic	0.3–0.8	$\geq$96	$\geq$95	$\geq$94
	Extra heavy, extremely heavy traffic	0.3–1.2	$\geq$96	$\geq$95	–
On the embankment	Light, medium and heavy traffic	0.8–1.5	$\geq$94	$\geq$94	$\geq$93
	Extra heavy, extremely heavy traffic	1.2–1.9	$\geq$94	$\geq$94	–
Under embankment	Light, medium and heavy traffic	$\leq$1.5	$\geq$93	$\geq$92	$\geq$90
	Extra heavy, extremely heavy traffic	$\leq$1.9			

Note 1. The compaction degree listed in the table is the compaction degree of the maximum dry density obtained via the heavy compaction test method of the current "Highway Geotechnical Test Regulations" (JTG E40) [1]

2. When asphalt concrete and cement concrete pavements are paved on third- and fourth-grade highways, the compaction degree of the road should be the standard of the second-grade highway

3. When the embankment adopts special fillers such as fly ash and industrial waste residue or is in a special arid or special humid area, on the premise of ensuring the requirements of the strength and elastic modulus of the subgrade, the standard of compaction can be reduced by 1–2% points

Table 6.5 Standard for quality control of hard stone compaction

Subgrade area	Depth below the surface of the road (m)	Thickness of paving layer (mm)	The biggest size (mm)	Dry density of compaction (kN/m^3)	Porosity (%)
On the embankment	0.80–1.50 (1.20–1.90)	$\leq$400	<2/ 3thickness	Determined by test	$\leq$23
Under the embankment	>1.50 (>1.90)	$\leq$600	<2/ 3thickness	Determined by test	$\leq$25

Note In the "depth below the bottom of the road surface" column, the values in parentheses are the depth ranges of the upper embankment and the lower embankment for extraheavy and extremely heavy traffic, respectively

Table 6.6 Standard for quality control of medium hard stone compaction

Subgrade area	Depth below the surface of the road (m)	Thickness of paving layer (mm)	The biggest size (mm)	Dry density of compaction (kN/m^3)	Porosity (%)
On the embankment	0.80–1.50 (1.20–1.90)	≤400	<2/3thickness	Determined by test	≤22
Under the embankment	>1.50 (>1.90)	≤500	<2/3thickness	Determined by test	≤24

Construction (JTG F10) [3] stipulates that the construction quality of rock-filled embankments should be based on the porosity of the stone after compaction. As shown in Table 6.7, the classification of stone filling is carried out according to the saturated compressive strength index of the stone according to Table 6.8.

The compaction test methods for soil subgrades include the sand filling method, the ring knife method, the drainage method (water bag method) or the nuclear density hygrometer method. When the nuclear density hygrometer method is used, calibration should be carried out first.

Table 6.7 Standard for quality control of soft stone compaction

Subgrade area	Depth below the surface of the road (m)	Thickness of paving layer (mm)	The biggest size (mm)	Dry density of compaction (kN/m^3)	Porosity (%)
On the embankment	0.80–1.50 (1.20–1.90)	≤300	<Thickness	Determined by test	≤20
Under the embankment	>1.50 (>1.90)	≤400	<Thickness	Determined by test	≤22

Note In the "depth below the bottom of the road surface" column, the values in parentheses are the depth ranges of the upper embankment and the lower embankment for extraheavy and extremely heavy traffic, respectively

Table 6.8 Classification of rocks

Rock types	Uniaxial saturation compressive strength (MPa)	Representative rock
Hard rock	≥60	1. Granite, diorite, basalt and other magmatic rocks
Medium-hard rock	30–60	2. Siliceous, iron cement conglomerate and sandstone, limestone, dolomite and other sedimentary rocks 3. Gneiss, quartzite, marble, SLATE, schist and other metamorphic rocks
Soft rock	5–30	1. Tuff and other extrusive rocks 2. Sedimentary rocks such as glutenite, argillaceous sandstone, argillaceous shale and mudstone 3. Mica schist or melite and other metamorphic rocks

6.2.2 Cutting Excavation

According to the specific conditions, the horizontal full-width excavation method, that is, excavating one or both ends of the entire section of the cutting forward along the longitudinal direction, should be adopted. Deep cuttings can also be divided into several steps and excavated at several different heights at the same time to increase the working line; the longitudinal channel excavation method can also be used; that is, the channel is dug longitudinally along the cutting and then widened to both sides. For deep cuttings with a large amount of excavation and a short construction period, the mixed excavation method of double-layer vertical and horizontal passages can also be adopted, and excavation can be carried out in the positive and negative directions in the vertical and horizontal directions at the same time to expand the construction surface. If the bottom surface of the cut surface is solid, it should be leveled and compacted without disturbing it as much as possible; if the soil quality is poor and the horizontal condition is poor, the side ditch should be deepened, the underground subsurface drain should be set up, and the surface layer should be loosened to a certain depth according to the design requirements of the pavement strength. The soil layer should be relayered and compacted or replaced and reinforced if necessary to ensure the strength and stability of the soil foundation at the bottom of the cutting to meet the specified standards. This is especially important for building durable pavements.

Soil Cutting

Soil-cutting excavations, according to the number of excavations and construction methods, can be divided into longitudinal full-width excavations and transverse channel excavations according to the excavation direction and can be divided into single-layer or double-layer excavations in height and vertical and horizontal excavations mixed, etc. (the above excavation directions are named according to the vertical and horizontal directions of the route).

Longitudinal full-width excavation excavates longitudinally forward along the route at one or both ends of the route, as shown in Fig. 6.4. The height of single-layer excavation is equal to the design depth of the cutting. When excavating, it is formed and advanced segment by segment, and the earth is sent out from the opposite direction. The height of single-layer longitudinal excavation is limited by the safety of manual operation and the effectiveness of mechanical operation. If the construction period is urgent, for deeper cuttings, the double-layer excavation method can be used; the upper layer is in front, the lower layer is after, and the upper layer is left on the construction surface of the lower layer. excavation and drainage channels.

Horizontal channel excavation first excavates the channel in the longitudinal direction of the cutting and then excavates it horizontally in sections at the same time, as shown in Fig. 6.5. This method can expand the construction surface, speed up the construction progress, and is used when excavating long and deep cuttings. The

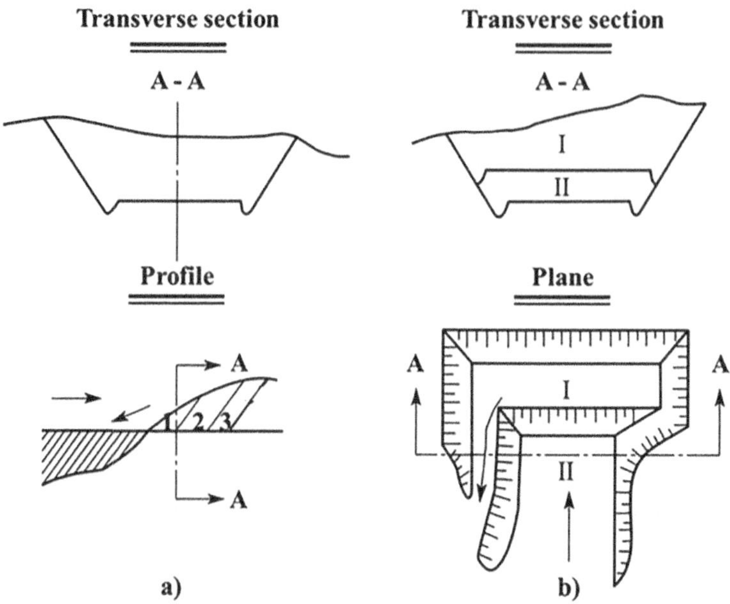

Fig. 6.4 Schematic diagram of longitudinal driving: **a** single layer; **b** double

construction can be divided into layers and sections, and the height and length of the sections depend on the construction method. This method has much work, but there are restrictions on earth-moving channels, and the interference of construction increases. It must be carefully arranged to prevent quality or safety accidents in chaos. In individual cases, to expand the construction surface and speed up the construction progress, the excavation of soil cuttings can also be considered to adopt a mixed excavation scheme of double-layer vertical and horizontal channels and simultaneously excavate multiple construction surfaces along the vertical and horizontal directions, as shown in Fig. 6.5b. The mixed excavation scheme is more intrusive and is generally limited to manual construction. For deep cuttings, it can be considered if the number of excavations is the largest and if the previous period is limited.

Stone Cutting

The excavation methods for stone cuttings mainly include the blasting method and loosening method, among which the blasting methods mainly include the steel needle method, deep hole blasting, shaped charge blasting, smooth surface/presplitting blasting and collapse blasting.

The excavation of stone cuttings should be based on different geology, different excavation sections, and different locations, and different excavation methods should be selected. For cuttings with depths of <4 m, line hole blasting should be adopted. Face blasting was used to control the slope rate, and loose blasting was used for the

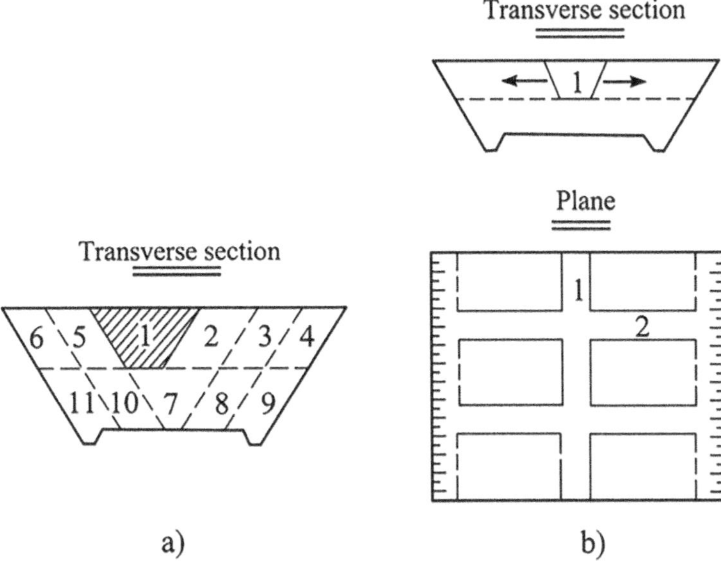

Fig. 6.5 Schematic diagram of lateral and mixed driving: **a** double lateral; **b** double mixing

other parts. Before excavation of the stonework, a blasting test should be carried out to select the best blasting parameters. After blasting, the slag is loaded by excavators and transported by dump trucks, and the road surface is repaired manually with mechanical brushing.

6.2.3 New Technology for Subgrade Construction

China is a country with a vast territory and complex and diverse geographical conditions. The technical problems faced in highway construction vary from place to place due to different specific conditions. Many areas face considerable technical difficulties in highway construction due to geographical and geological constraints. In recent years, with the continuous efforts of the majority of builders, various new technologies, new processes and new materials have emerged in the construction of highway engineering, and many technical problems have been overcome. Moreover, the national standard system of highway construction and acceptance in China is also in the process of continuous updating and improvement. This section introduces the treatment technologies involved in the widening and reconstruction of lightweight embankments and roads.

Subgrade Filler, Compaction, and Reinforcement

In China, each part of the subgrade filler has a corresponding specified value. When the subgrade filler fails to meet the specified minimum strength requirements, it should be mixed with coarse aggregates, replaced, or treated with stable materials such as lime, and when high-grade roads are paved on other grades of roads, the specified values for expressways and grade one roads should also be used.

In current subgrade construction, large-tonnage road rollers are generally used, and the rolling effect has been significantly improved, which has improved the compaction of the subgrade soil.

With the advancement of subgrade construction technology, the treatment technology for special subgrades is also becoming more mature and perfect. Construction techniques for soft soil foundations include lightweight embankments, geosynthetic reinforcement, CFG piles, and thin-walled tubular piles.

(1)   Light embankment

Lightweight embankment refers mainly to a subgrade filled with light (fly ash) or ultralight material (Expanded Polystyrene, EPS for English abbreviation) [6]. Fly ash, which is a waste of coal-fired power plants, can be used for subgrade construction to reduce land occupation and benefit the environment. Moreover, it has the characteristics of low self-weight and high strength, and the strength of the mixture increases with time, compactability, fast consolidation, and low cost. It can be used in soft soil foundation road sections, which can increase the height of subgrade filling and reduce subgrade and foundation settlement. Since the research and application of the subgrade filling of the Shanghai–Nanjing Expressway in the 1980s, technical specifications for design and construction have been established, which are not described here.

The volume weight of the ultralight material EPS is approximately 0.18–0.4 kN/m^3, which is approximately 1/50–1/70 of the weight of ordinary soil at 14–20 kN/m^3. The water absorption rate of EPS is very small, and the heat resistance and water resistance are very good. It has a certain strength, and the compressive strength is 100–300 kPa. Typically, the pressure of the overlying embankment on the subgrade is <100 kPa. Therefore, EPS can be used as embankment filler to reduce the weight of the embankment, reduce the settlement of the embankment, and ensure the stability of the embankment. When EPS is used as embankment filler, construction is very convenient. The size of EPS blocks can be produced according to need. The sizes of EPS blocks are usually 3 m long, 1 m wide, and 0.5 m thick; similar to hand-held building blocks, there is no need for large machinery. The strength of EPS itself is sufficient to meet the requirements of embankment load and slope stability. In application, the EPS filling height is generally approximately 5 m, and the filling height can reach 20 m. In addition, when EPS is used as embankment filler, the slopes on both sides of EPS are edged with soil, and the stability of the embankment slope depends on the stability of the edged soil, whereas the stability of the edged soil can be determined via conventional soil mechanics. The method of slope stability is determined. Since EPS itself is a block and has strong cohesion, the embankment using EPS filler will make the embankment slope more stable.

In EPS embankment construction, the EPS must be laid flat. To ensure good drainage performance of the EPS-filled subgrade, a permeable sand layer should be laid at the bottom of the EPS at the bottom, and the flatness of the sand layer and the EPS paving layer should be strictly controlled. No space should be left between the blocks, and the blocks should be closely arranged. The EPS blocks should be combined with adhesive materials or seam nails to prevent mutual dislocation and displacement of the EPS blocks during the embankment filling process. Gaps or height differences must be created, which must be adjusted with nonshrinkage cement mortar. The disadvantage of the use of EPS as an ultralight filler for embankments is its poor weathering resistance, impact resistance and chemical properties. However, when it is used as an embankment filler, it is buried in the soil, and it is less affected by ultraviolet rays and ages slowly, so its strength is low. Deterioration is small. Moreover, EPS is not suitable for use as an ultralight filler for embankments in surface flooded areas to avoid surface flood buoyancy lifting the EPS embankment, resulting in embankment damage. In addition, due to the high cost of EPS, it is mainly used in the filling of emergency repair projects in foreign countries, as well as projects that have restrictions on the overlying pressure acting on the structure.

(2)   Geosynthetic reinforcement

For soft soil foundations with shallow layers (generally <3 m in thickness), geotextiles can be laid on the surface first, and then embankments can be filled. method. The thickness of the soft soil layer is 3–5 m, the combination of a geotextile and a sand cushion can be used for treatment, and the thickness of the drainage sand cushion can be reduced from 50 to 30 cm. There are also multilayer geotextiles laid between the embankment and the surface, and the high tensile strength of the material is used to overcome the sliding deformation of the foundation to maintain stability. By controlling the filling rate and cooperating with overload preloading, the foundation is quickly consolidated [7].

(3)   Reinforcement of CFG piles and thin-walled tubular piles

The term "CFG pile" refers to "cement fly ash gravel" in English. It is a variable-strength pile made by various pile-forming machines that mixes gravel, stone chips, sand, and fly ash with cement and water. By adjusting the cement content and ratio, its strength grade varies between C15 and C25, which is a type of pile between rigid piles and flexible piles. The CFG pile and the soil between the piles work together to form the CFG pile composite foundation through the cushion layer, and the engineering design can be carried out according to the properties and calculation of the composite foundation. CFG piles generally do not need reinforcement calculations and can also use industrial waste fly ash and stone chips as admixtures to further reduce project costs.

The following construction techniques should be selected for CFG piles according to the site conditions: (1) Long spiral boreholes are poured into piles, which are suitable for clay soils, silty soils, plain fill soils, and pile soils with medium density above the groundwater level. (2) Long screw drilling, in-pipe pump pressure mixtures poured into piles, are suitable for cohesive soil, silt, sandy soil, and sites with strict requirements for noise or mud pollution. (3) The vibrating immersed tube is cast into piles, which is suitable for silt, cohesive soil and plain fill foundations.

Cement concrete upper thin-walled pipe piles are hollow thin-walled pipe piles made of cement concrete and steel bars by centrifugal molding, and the piles are driven into the foundation by a piling machine to form a pile foundation platform. Pretensioned prestressed high-strength concrete pipe piles are called PHC piles, pretensioned prestressed concrete upper pipe piles are called PC piles, pretensioned prestressed high-strength concrete thin-walled pipe piles are called PTC piles, and cast-in-place concrete thin-walled piles are called PTC piles. Pipe piles are called PCC piles. Cement concrete thin-walled pipe piles can be used in the same building foundation, and pipe piles of different diameters can be used, which makes it easy to solve the problem of pile layout and can give full play to the bearing capacity of each pile. The construction conditions are limited; the quality of the pile is reliable, and the length and quality of the pile body can be monitored directly after the pile is driven; the pile body has good hammer resistance, good crack resistance, and strong penetration; it is 1/3–2/3 of the steel pile and saves steel, fast construction, and less settlement after construction.

Subgrade Drainage

The design of road drainage should include the following two aspects: one is to consider how to reduce the impact of groundwater and farmland drainage and irrigation on the stability and strength of the subgrade, generally referred to as the first type of drainage; water is quickly drained out of the subgrade, minimizing the impact of rainwater on the quality of the subgrade and pavement and reducing damage to the subgrade, pavement structure and performance due to poor drainage of road surface water or infiltration of surface water. Class II drainage.

The first type of drainage design usually adopts methods such as appropriately increasing the minimum filling height of the subgrade or setting a waterproof cushion at the bottom of the subgrade.

During construction, temporary drainage side ditches are generally considered before construction to remove surface water and reduce groundwater. Simultaneously, low-dose lime was added to the bottom of the subgrade for treatment, and a 40 cm thick stable layer was set up. Using this series of measures can have a very good effect.

Subgrade Protection

The construction of the subgrade changes the natural balance of the stratum, and the subgrade is exposed to space and constantly eroded, so various types of protection are needed .

(1) Slope protection

The purpose of slope protection is to prevent erosion of surface water, weathering and peeling of slope rock and soil, and coordination with the environment. In recent years, with an emphasis on environmental protection, the slopes of high-grade highways have been protected mostly by grass; when the slope is high, a masonry frame has been used for single protection. Owing to the drought and water shortage in the west, the choice of grass protection type is very important. At present, most lawn planting belts are used; that is, grass seeds, fertilizers and soil are evenly mixed and wrapped in geotextiles (such as nonwoven fabrics). After the grass grows to fix the soil, the nonwoven fibers rot naturally without polluting the environment, and the effect is very good.

(2) Scour protection

To protect a certain slope along the river road from scouring, direct protection is still used. The traditional masonry, riprap, wire gabion, retaining wall, etc., have improved. High-strength geogrids are used instead of wires as gabions. A protective panel made of ester or polyurethane geotextile concrete slope protection molds protects slopes impacted by water waves and can adapt to the uneven settlement of soil.

(3) Support protection

The retaining walls are still mainly used for support and protection. Stone-built gravity retaining walls are mostly used when diced stone is abundant, the wall height is low, and the foundation is good. Retaining walls and slab–column retaining walls have relatively reasonable stresses and small masonry volumes and have also been widely used in the protection of highway subgrades. The height of the retaining wall used for stacking can be easily adjusted, and the wall can be assembled from prefabricated components. It is a special retaining wall.

Combined Treatment Technology for New and Old Subgrades

Since the beginning of the twenty-first century, many roads in China have been saturated with traffic, and widening and reconstruction have been implemented. The key technical problem to be solved is the coordinated deformation of the joint between the new and old subgrades. Under different conditions, the composition of the uncoordinated deformation of the old and new subgrades is different. To ensure the stability of the subgrade, measures must be taken to control the uncoordinated deformation of the subgrade. According to the location and treatment mechanism of the treatment measures, the control technology of uncoordinated deformation can be divided into four categories: internal treatment of the pavement, internal treatment of the subgrade, external treatment and comprehensive treatment (see Table 6.9).

If divided according to the main sources of uncoordinated deformation at the junction of new and old subgrades, the treatment technologies in Table 6.9 can be divided

Table 6.9 Preliminary classification of treatment technology at the junction of the old and new subgrades

Treatment technology of new and old subgrade junction	Treatment technology of new and old subgrade junction	Increasing thickness
		Improve anti-deformation ability (stiffening, setting mesh)
	Internal treatment of subgrade	Joint surface treatment
		Packing and compaction control
		Subgrade reinforcement
		Lightweight embankment
	External treatment	Lightweight embankment
		Foundation treatment
		Retaining structure
	The comprehensive treatment	Set the divider
		Improve drainage system
		Transitional road surface
		Internal and external comprehensive treatment

into the following: ground treatment technologies for the poor geological conditions at the junction of new and old subgrades; slope treatment and joint reinforcement technology; control the degree of subgrade packing and compaction, lightweight road piles and other measures for the excessive compression deformation of the subgrade itself; if the uncoordinated deformation of the new and old subgrade joints is caused by the above, if the factors are combined, comprehensive treatment techniques should be adopted, as shown in Table 6.10.

In actual widening and reconstruction projects, different treatment methods are often used according to specific project characteristics, and a variety of treatment techniques are sometimes used comprehensively. The design and construction of the junction of the old and new subgrades is a very important part of the entire reconstruction project, which requires careful design and construction to ensure the quality of the project. For specific requirements, please refer to the current technical specifications for highway subgrade design and construction, which will not be described here.

Table 6.10 Treatment techniques and applicable conditions for discordant deformation sources

The main source of discordant deformation at the junction of old and new subgrade	Joint treatment technology	Suitable conditions
Consolidation settlement of foundation under the action of new subgrade	The joint foundation is treated by the methods of replacing fill, squeezing silt with riprap, composite foundation and drainage consolidation	Embankment widening under unfavorable geological conditions, high embankment filling, etc.
The joint strength of new and old subgrade is insufficient	The slope of the old road is covered with soil, the step is excavated, and the joint is set with geogrid	The strength of the slope soil of the old road is low due to natural weathering and it is difficult to join the old and new subgrade
Self-compression deformation of new and old subgrade	The new subgrade packing is preferred to improve the compaction degree, and the new subgrade adopts the lightweight embankment of ash and EPS	The subgrade with better geological conditions is widened
The composition of the above factors	The above treatment technology is comprehensively applicable, considering the setting of retaining wall, pavement auxiliary treatment technology and the improvement of drainage system	All kinds of poor foundation, subgrade and joint surface conditions

6.3 Asphalt Pavement Construction

6.3.1 Preparation of Materials and Equipment

Asphalt Material

The asphalt material should be heated with heat transfer oil. The heating temperature of ordinary asphalt should meet the requirements of the main technical specification for highway asphalt pavement construction (JTG F40) [4] (Table 6.11) conveyed to the mixer.

Aggregate

To ensure the cleanness of the aggregate, the ground of the aggregate yard should be hardened with cement concrete, and the roads entering the mixing plant and the aggregate yard should also be hardened with cement concrete. To ensure that the

Table 6.11 Determining the appropriate temperature for the mixing and compaction of an asphalt mixture

Viscosity	Viscosity of asphalt binder suitable for mixing	Viscosity of asphalt binder suitable for compaction	Determination method
Apparent viscosity	(0.17 ± 0.02) Pa s	(0.28 ± 0.03) Pa s	T 0625
Kinematic viscosity	(170 ± 20) mm^2/s	(280 ± 30) mm^2/s	T 0619
Saybolt viscosity	(85 ± 10) s	(140 ± 15) s	T 0623

aggregates are not mixed with each other, the aggregates of different specifications should be isolated [10]. The aggregate yard should be set up with a shed, and the fine aggregate should be covered with oil cloth to prevent the aggregate from becoming wet. The technical requirements for aggregates comply with the requirements of the "Technical Specifications for Construction of Highway Asphalt Pavement" (JTG F40) [4]. The moisture content of the aggregate should not exceed 1% when it is fed into the mixing equipment.

Asphalt Mixing Equipment

The mixing equipment of the asphalt mixture should adopt an automatic control batch mixer, and the mixer should meet the following requirements:

(1) Automatic control. Automatically controlled mixing equipment should be able to use computers and other equipment to easily adjust the mixing ratio and be equipped with thermometers and finished product storage bins with heat preservation functions and secondary dust removal equipment. The mixing equipment should be controlled by a computer, and the heating temperature of the aggregate and asphalt, the mixing temperature of the mixture, the amount of material, and the production machinery of each plate of the mixture should be printed disk by disk. The production of the mixing equipment should match the production schedule. After the installation is completed, trial mixing should be carried out according to the approved mixing ratio until it meets the requirements.

(2) Dust collector. The mixer should be equipped with a dust collector, and its structure should be able to digest all or part of the materials to be collected according to regulations and prevent harmful dust from escaping into the air. To prevent dust from being released into the air, it is necessary to cover the dust filter with a dust seal.

(3) The layout of the mixing site should be far from the residential area, and the distance should not be <1 km.

6.3.2 Mixing of Asphalt Mixtures

(1) Asphalt should be heated with heat-conducting oil, the heating temperature of the asphalt and aggregate should be adjusted so that the temperature of the mixed asphalt mixture can meet the requirements [9], and the aggregate temperature should be 10–20 °C higher than the asphalt temperature. The heating temperature of the mixture and the discharge temperature of the asphalt mixture. When the ex-factory temperature of the mixture is too high, which affects the adhesion between the asphalt and the aggregate, the mixture should not be used, and the paved asphalt pavement should also be removed.

(2) The mixing time is determined by trial mixing. The aggregate particles used must be evenly coated with the asphalt binder, and the asphalt mixture must be mixed evenly. The mixing time of each pot of the batch mixer should be 30–50 s (the dry mixing time should not be <5 s). The sieve holes of the vibrating screen used for the secondary screening of hot ore should be selected according to the grading of the ore, and the installation angle should be determined by tests according to the sieve ability and vibration capacity of the material.

(3) The asphalt mixture mixed by the mixing plant should be uniform, without whitening, agglomeration or severe separation of coarse and fine materials. It should not be used when it does not meet the requirements and should be adjusted in time. When the mixed hot mix asphalt mixture is not paved immediately, it can be put into the finished product storage silo for storage. After the finished hot concrete is stored in the storage silo, the temperature drop should not exceed 5 °C. The storage time of the storage silo should not exceed 24 h and should not exceed 48 h at most.

(4) The control room of the mixing building should print the amount and temperature of the asphalt and various mineral materials disk by plate and regularly check the metering and temperature measuring system of the mixing building; mixers without material metering and automatic temperature metering devices should not be used. The total amount of mixing is used to check the error of the mixing ratio of the mineral aggregate and the oil-to-stone ratio of the asphalt mixture every day.

(5) The asphalt mixture should meet the requirements of the approved site mix ratio and should be within the allowable deviation range of the target value, and the allowable deviation of the target value of aggregate gradation should meet the requirements of Table 6.12.

6.3.3 Transportation of Asphalt Mixtures

(1) The car transporting asphalt mixture should have a tight, clean and smooth metal floor, and the floor should be coated with a thin layer of oil–water mixture (the ratio of diesel and water can be 1:3) to prevent the mixture from sticking to the floor, but no excess liquid should accumulate on the bottom of the carriage.

Table 6.12 Allowable deviation of the aggregate grading target value

Item		Inspection frequency and evaluation method of single point inspection	Quality requirements or allowable deviations		Test method
			Highway, first class highway	Other grade highway	
Ore grading (sieve)	0.075 mm	Disk by disk detection	±2% (2%)	–	Calculation and acquisition data calculation
	≤2.36 mm		±5% (4%)	–	
	≥4.75 mm		±6% (5%)	–	
	0.075 mm	Check disk by disk, summarize once a day, and take the average value for evaluation	±1%	–	JTG F40 Appendix G Gross Inspection
	≤2.36 mm		±2%	–	
	≥4.75 mm		±2%	–	
	0.075 mm	Each mixer is evaluated by the average value of 2 samples 1–2 times a day	±2% (2%)	±2%	T 0725 comparison of extraction and screening with standard grading
	≤2.36 mm		±5% (3%)	±6%	
	≥4.75 mm		±6% (4%)	±7%	

The use of petroleum-derived agents for the coating of automobile bodies is not allowed. Before loading and unloading, the vehicle floor should drain the water. Every car should have a canopy made of canvas, quilt and other materials, the size of which should be able to protect the mixture from the weather. fall. To transport the mixture to the pavement construction site at the specified temperature, the vehicle floor should be insulated if necessary, and the canvas should be fastened.

(2) All drivers should be trained before construction, and the maintenance of the vehicle should be strengthened to avoid damage to the cooling of the mixture caused by the breakdown of the vehicle during transportation; the vehicle should move back and forth during loading to avoid separation of the mixture; and the transportation vehicle should be on the paver. During the unloading process, the truck should be in the neutral gear and pushed forward by the paver to ensure the flatness of the paving layer.

(3) The transport capacity of the asphalt mixture transporter should be greater than the mixing capacity or the paving capacity. During the construction process, there should be a transporter in front of the paver waiting for unloading. For expressways and first-class highways, there should not be fewer than 5 trucks waiting for unloading at the construction site when paving begins. After the asphalt mixture is delivered to the paving site, it should be received with the waybill, and the mixing quality should be checked. Mixtures that do not meet

the temperature requirements of JTG F40 or that have formed agglomerates and have been wetted by rain should not be paved on the road.

6.3.4 Paving of Asphalt Mixtures

Paving of Asphalt Mixtures

The hot mix asphalt mixture should be paved with a paver, and a crawler paver should be used when paving the modified asphalt mixture or SMA on the road surface sprayed with sticky oil. When paving asphalt mixtures for highways and first-class highways, the paving width of one paver should not exceed 6 m, and two or more pavers should be staggered by 10 on one-way two-lane or three-lane highways. – 20 m is paved synchronously in an echelon manner. Pavers must be continuous, slow, and uninterrupted to improve flatness and reduce mixture segregation. The paving speed should be controlled within the range of 2–6 m/min, and the modified asphalt mixture and SMA should be controlled within the range of 1–3 m/min. The paver should adopt the automatic levelling method.

Temperature Control

The paving temperature of the asphalt mixture should meet the requirements of JTG F40 [4] (Table 6.13) and should be selected according to the asphalt grade, viscosity, air temperature and thickness of the paving layer.

The construction temperature of the polymer-modified asphalt mixture is selected according to practical experience and with reference to Table 6.14. Typically, the construction temperature of ordinary asphalt mixtures should be increased by 10–20 °C. For the modified asphalt mixture produced by the direct injection method of cold latex, the drying temperature of the aggregate should be further increased.

6.3.5 Rolling of Asphalt Mixtures

(1) A reasonable combination of rollers and rolling steps should be selected to achieve the best results. Asphalt mixture compaction should involve the combination of a steel drum static roller and a tire roller or vibratory roller, and a tire roller should not be used for initial pressure to ensure a smooth surface. The number of road rollers should be determined according to the road width, etc.

(2) The maximum thickness of the compacted layer of asphalt concrete should not be >100 mm, and the thickness of the compacted layer of the asphalt-stabilized gravel mixture should not be >120 mm; however, when a high-power road roller

Table 6.13 Construction temperature of the hot mix asphalt mixture

Construction process		The labeling of petroleum bitumen			
		50#	70#	90#	110#
Asphalt heating temperature		160–170	155–165	150–160	145–155
Mineral heating temperature	Batch mixer	The aggregate heating temperature is 10–30 higher than that of asphalt			
	Continuous mixer	The heating temperature of mineral material is 5–10 higher than that of asphalt			
Discharge temperature of asphalt mixture		150–170	145–165	140–160	135–155
Mixture storage bin storage temperature		The temperature decrease during material storage shall not exceed 10 °C			
Discard temperature of mixture, >		200	195	190	185
Transport to the site temperature, ≥		150	145	140	135
Mixture spreading temperature, ≥	Normal construction	140	135	130	125
	Low temperature construction	160	150	140	135
The internal temperature of the mixture at which the rolling begins, ≥	Normal construction	135	130	125	120
	Low temperature construction	150	145	140	135
Surface temperature at the end of rolling, ≥	Steel wheel roller	80	70	65	60
	Tyre roller	85	80	75	70
	Vibratory roller	75	70	60	55
Pavement temperature for open traffic, ≤		50	50	50	45

Note 1. The construction temperature of the asphalt mixture can be measured by using a plug-in digital thermometer with a metal probe. The surface temperature can be measured by a contact thermometer. When an infrared thermometer is used to measure the surface temperature, it should be calibrated.

2. The construction temperatures of the Nos. 130, 160 and 30 asphalts not listed in the table are determined via tests

is used and the degree of compaction can be proven by tests, it can be increased to 150 mm.

(3) Asphalt pavement construction should be equipped with a sufficient number of road rollers, and a reasonable combination of road rollers and the initial pressure, recompression, and final pressure (including forming) rolling steps should be selected to achieve the best rolling effect. The number of road rollers for laying two-lane asphalt pavement on expressways should not be <5. When the construction temperature is low, the wind is strong, and the rolling layer is

Table 6.14 Normal operating temperature range of polymer-modified asphalt mixtures (°C)

Process	Polymer modified asphalt varieties		
	SBS classes	SBR latex class	EVA, PE class
Asphalt heating temperature	160–165		
Field production temperature of modified asphalt	165–170	–	165–170
Heating temperature of finished modified asphalt, ≤	175	–	175
Aggregate heating temperature	190–220	200–210	185–195
Factory temperature of modified asphalt SMA mixture	170–185	160–180	165–180
Maximum temperature of mixture (waste temperature)	195		
Mixture storage temperature	Reduce by <10 after mixing		
Paving temperature, ≥	160		
Initial pressure initial temperature, ≥	150		
Surface temperature at the end of crushing	90		
Road gauge temperature when open to traffic, ≤	50		

Note When asphalts are modified with nonlisted polymers or natural asphalts, the operating temperature is determined via tests

Table 6.15 Roller rolling speed (km/h)

Type of roller	First compaction		Second compaction		Final compaction	
	Appropriate	Maximum	Appropriate	Maximum	Appropriate	Maximum
Steel wheel roller	2–3	4	3–5	6	3–6	6
Tyre roller	2–3	4	3–4	6	4–6	8
Vibratory roller	2–3 (static pressure or vibration)	3 (static pressure or vibration)	3–4.5 (vibration)	5 (vibration)	3–6 (static pressure)	6 (static pressure)

thin, the number of road rollers should be appropriately increased. The rolling speed of the road roller should meet the requirements in Table 6.15.

6.3.6 Treatment of Construction Joints

When two or more pavers are used for echelon operation, the paved mixture should be left 10–20 cm wide and not rolled temporarily, as the rear elevation datum, and there should be a 5–10 cm dedicated paving layer overlapping for thermal bonding.

The seam form is rolled across the seam at the end to eliminate the seam. The upper and lower longitudinal seams should be staggered by more than 15 cm.

The horizontal construction joints are flat joints. A 3 m ruler is placed in the longitudinal direction so that the ruler at the end of the paving section is cantilevered, the joint position at the point where the paving layer and the ruler are out of contact is determined, and the ruler is cut with a sawing machine and removed. When the ruler continues to pave, the mortar left by the joint sawing should be scrubbed clean and coated with sticky asphalt, and the paver ironing plate should start paving after the joint. The span is gradually moved toward the newly paved surface. The horizontal construction joint should be 20 cm away from the expansion joint of the bridge and should not be set at the expansion joint to ensure the smoothness of the road surface on both sides of the expansion joint.

6.4 Cement Concrete Pavement Construction

6.4.1 Construction Preparation

Equipment Requirements

Generally, under the technical level of construction, the construction of different grades of highway cement concrete pavement should meet the requirements of Table 6.16.

Set Template

The formwork is made of rigid form or other materials and meets the requirements of the design of the road surface, vertical and horizontal, to ensure that the formwork connection is firm and reliable and that the support is stable so that it can withstand the impact and vibration of tamping and finishing equipment when pouring concrete. No displacement occurs, and the height of the formwork is the same as the thickness of the concrete pavement.

The formwork at the construction joint should be set out and drilled according to the design position of the dowel rod or tie rod, and there should be a firm splicing device at the formwork joint, which is easy to assemble and disassemble.

Dowel Bar Set

The dowel bars set at the transverse shrinkage joints and expansion joints should be parallel to the centerline and the surface of the road surface. A layer of asphalt was added 5 cm to the half length of the dowel bar to ensure the free expansion and

Table 6.16 Equipment requirements for the construction of different grades of highway cement concrete pavement

Paving process machinery and equipment	Highway	First class road	Second class road	third class road	Fourth class road
Slide form pavers	√	√	√	●	○
Three roller shaft unit	○	○	√	√	√
Small machine	×	×	●	√	√
Roller compacted concrete	●	●	√	√	√
Automatic computer control of forced mixing building (station)	√	√	√	●	○
Forced mixing building (station)	×	○	●	√	√

Note 1. Symbol meaning: √ shall be used; ● conditional use; ○ not recommended; ×Shall not be used

2. Volume measurements and small self-falling drum mixers are not allowed on highways of all grades, and manual control of water addition is strictly prohibited

contraction of the panel. A prefabricated cover was added to the asphalt-coated end of the dowel bar at the expansion joint, a 30 mm gap was left, which was filled with yarn or foam.

The tie rod must be installed prior to concrete paving, either with a tie rod vibrator at the joint edge or with a tie rod automatic rod threader on the concrete paver.

6.4.2   Construction of Cement Concrete Pavement

Mixing of the Concrete Mixtures

The production capacity of the total mixing equipment in the mixing plant is required to ensure that the actual paving capacity is met, and the number and type of mixing buildings required are determined according to the total mixing capacity. The minimum configuration capacity of the mixing buildings for different paving methods of concrete pavement is shown in Table 6.17.

Table 6.17 Minimum configuration capacity of mixing buildings with different paving modes of concrete pavement

Paving width	Paving width	Roller compacted concrete	Three roller shaft unit paving	Small unit paving
One lane 3.75–4.5 m	$\geq$150	$\geq$100	$\geq$75	$\geq$50
Two lanes 7.5–9 m	$\geq$300	$\geq$200	$\geq$100	$\geq$75
The whole width $\geq$ 12.5 m	$\geq$400	$\geq$300	–	–

Each mixing plant must be calibrated and mixed before it is put into production. The calibration should be recalibrated after the validity period of calibration expires or after the mixing building is relocated and installed. During construction, the measurement accuracy of the mixing building should be checked every 15 days. When a mixing plant with an automatic computer control system is used, it should use automatic batching production and print the concrete batching statistics and deviations corresponding to the pavement paving pile number every day (week, ten, and month) as needed.

During the stirring process, the inspection and control of the mixed substances should comply with the provisions of Table 6.18. During construction in low-temperature or high-temperature weather, the discharge temperature of the mixture should be controlled at 10–35 °C, and the temperature of the raw materials, the temperature of the mixture, the slump loss rate and the setting time should be measured.

Transportation of Concrete Mixtures

The transportation of cement concrete materials should be based on the construction progress, transportation volume, transportation distance and road conditions, and the total number of models and vehicles should be selected. The total transportation capacity should be slightly greater than the total mixing capacity to ensure that the freshly mixed cement concrete is transported to the paving site within the specified time. The maximum allowable time for concrete mixtures of different paving processes from the discharge of the mixer to the completion of transportation and paving should meet the requirements of Table 6.19. If not satisfactory, the test should be passed, and the dosage of retarder or plasticizer should be increased.

Paving of the Cement-Concrete Mixture

When the cement concrete is being laid, the cement concrete dumped in the base layer or the paving box should be evenly filled within the scope of the formwork according to the paving thickness. There are three main types of equipment:

Table 6.18 Quality inspection items and frequency of concrete mixtures

Check the project	Check the frequency	
	Highway, first class highway	Other grade highway
Water cement ratio and stability	Every 5000 m^3 sampling once, there are changes at any time to measure	Every 5000 m^3 sampling once, there are changes at any time to measure
Slump and its loss rate	Each shift is measured 3 times, and changes are measured at any time	Each shift is measured 3 times, and changes are measured at any time
Vibrational viscosity coefficient	Test mix, raw material and mix ratio when there is change	Test mix, raw material and mix ratio when there is change
Fiber volume ratio	Each shift is measured 2 times, there are changes at any time to measure	Each shift is measured 1 times, there are changes at any time to measure
Air content	Each shift is measured 2 times, there are frost resistance requirements not <3 times	Each shift is measured 1 times, there are frost resistance requirements not <3 times
Exudation rate	Each shift is measured 2 times	Each shift is measured 2 times
The apparent density	Each shift is measured 1 times	Each shift is measured 1 times
Temperature, setting time, hydration calorific value	Construction in winter and summer. Test at least 1–2 times per shift	Construction in winter and summer. Test at least 1–2 times per shift
Improve the VC value	Each shift is measured 3 times, and changes are measured at any time	Each shift is measured 3 times, and changes are measured at any time
Segregation	Observe	Observe
Degree of compaction, coefficient of loosening	Each shift is measured 3 times, and changes are measured at any time	Each shift is measured 3 times, and changes are measured at any time

Table 6.19 Maximum time allowed for delivery of the concrete mixture to the site

Construction temperature (°C)	Sliding mode paving (h)	Three roller shaft unit, Small unit paving (h)	RCC placement (h)
5–9	1.5	1.20	1.0
10–19	1.25	1.0	0.8
20–29	1.0	0.75	0.6
30–35	0.75	0.40	0.4

(1) Slipform paver

For the construction of expressways and first-class highways, a slipform paver that can pave 2–3 lane widths (7.5–12.5 m) at a time should be selected; the road design width. The paving of a hard shoulder should be equipped with medium and small multifunctional slipform pavers, and the curb should be paved as a whole at one

Table 6.20 Basic technical parameters of slide-mode pavers

Item	Power of engine (kW) $\geq$	Paving width (m)	Maximum spread thickness (mm) $\leq$	Range of spreading speed (m/min)	Maximum idle speed (m/min) $\leq$	Maximum walking speed (m/min) $\leq$	Number of tracks
Three-lane slide paver	200	12.5–16.0	500	0.75–0.30	5.0	15	4
Two-lane slide paver	150	3.6–9.7	500	0.75–0.30	5.0	18	2–4
Multifunction single lane slide paver	70	2.5–6.0	400 Maximum height of guardrail $\leq$1900	0.75–0.30	9.0	15	2–4
Small kerb slide paver	60	0.5–2.5	450	0.75–0.20	9.0	10	2–3

Table 6.21 Three roller shaft levelling machine main technical parameters

Shaft diameter (mm)	Axial speed (r/min)	Axial length (m)	Shaft mass (kg/m)	Walking speed (m/min)	Leveling wheelbase (mm)	Vibration power (kW)	Driving power (kW)	Suitable leveling pavement thickness (mm)
168	300	5–9	65 $\pm$ 0.5	13.5	504	7.5	6	200–260
219	380	5–12	77 $\pm$ 0.7	13.5	657	17	9	160–240

time. The selection of basic technical parameters of the slipform pair is shown in Table 6.20.

(2) Trimmer-finisher

The main technical parameters of the three-roller leveler are shown in Table 6.21. Rollers with a diameter of 168 mm should be used for plates with a thickness of more than 200 mm; rollers with a diameter of 219 mm can be used for bridge deck pavement or pavement with a small thickness. The shaft length should be 600–1200 mm longer than the road width.

Tamping of the Cement-Concrete Mixture

The position of the lower edge of the vibrator of the slipform paver should be above the lowest point of the extrusion plate, the vibrator should be evenly arranged, and the spacing should be 300–450 mm; the distance between most edge vibrators on both sides and the paving edge should not be >200 mm to ensure that the cement concrete vibrating in the entire range is compact and uniform. The forward inclination angle of the extrusion bottom plate should be set to approximately 3°. The position of the pulping and ramming plate should be 5–10 mm below the front edge of the extrusion bottom plate, and the overlaid height of the two edges should be determined according to the consistency of the mixture, which should be 3–8 mm. The front edge of the flattened beam should be adjusted to the same elevation as the rear edge of the extruded plate, and the rear edge of the flattened beam should be 1–2 mm lower than the rear edge of the extruded bottom plate and should have the same elevation as the road surface. When the trimmer-finisher is laying concrete panels, the vibrator group is intermittently inserted into the vibrating time. The row vibrator is continuously towed and vibrated in real time, and the operating speed should be controlled within 4 m/min. The row-type vibrator should vibrate at a constant speed, slowly, continuously and uninterruptedly. The operating speed is subject to the condition that the surface of the mixture does not contain coarse aggregates and that the liquefied surface no longer contains bubbles and no longer oozes the cement slurry. For single-lane paved concrete pavement, tie rods should be inserted into the reserved holes of the side formwork according to the design requirements. When paving two-lane road pavement for the second time, in addition to the tie rods being inserted into the side formwork holes, they should also be inserted into the middle longitudinal joint. The tie rod is used to insert the tie rod at 1/2 plate thickness, and the distance of each movement of the insertion machine should be the same as the distance between the tie rods.

Finishing, Refurbishment, Sawing and Maintenance

The vibrated and compacted cement concrete surface should keep its road crown accurate, and its flatness should meet the requirements. Before surface refurbishment, the edges and seams should be cleaned, the slime should be removed, and the missing edges and missing corners should be repaired.

When the concrete has hardened enough to withstand the sawing equipment, the sawing operation can begin. All sawdust and debris should be thoroughly removed immediately after the kerf work is completed.

After the surface of the concrete slab is finished, wet curing and plastic film curing should be adopted for 14–21 days.

Opening to Traffic

When the concrete slab reaches the design strength, it is allowed to open to traffic. When it is necessary to open the traffic in advance in special circumstances, the strength of the test block cured under the same conditions as the concrete panel should be measured according to the prescribed test method, which should reach more than 80% of the design strength, and the vehicle load should not be greater than the design load. Before opening to traffic, the pavement should be swept, and all the seams should be closed.

Discussion, Questions and Exercises

(1) What are the primary factors that influence the stability and strength of the road subgrade?
(2) Explain the role of drainage in subgrade construction. How can poor drainage affect the long-term performance of roads?
(3) Explain the basic process of the asphalt mix design. What factors influence the selection of asphalt binder and aggregate?
(4) What are the different types of asphalt mixtures, and how are they used in different layers of the pavement structure?
(5) Explain the process of preparing and placing concrete for pavement construction. What are the key quality control measures that need to be taken during construction?
(6) The role of curing in cement concrete pavement construction and its effect on the durability and strength of the pavement are discussed.

References

1. Ministry of Transport of the People's Republic of China. Test Methods of Soils for Highway Engineering (JTG 3430-2020). China Communications Press, Beijing, 2020
2. Ministry of Transport of the People's Republic of China. Technical Standard of Highway Engineering (JTG B01-2014). China Communications Press, Beijing, 2014
3. Ministry of Transport of the People's Republic of China. Technical Specifications for Construction of Highway Subgrades (JTG/T 3610-2019). China Communications Press, Beijing, 2019
4. Ministry of Transport of the People's Republic of China. Technical Specifications for Construction of Highway Asphalt Pavements (JTG F40-2023). China Communications Press, Beijing, 2023
5. Ministry of Transport of the People's Republic of China. Code for Design of Highway Subgrade (JTG D30-2015). China Communications Press, Beijing, 2015
6. Stark TD, Arellano D, Horvath JS, Leshchinsky D. Guideline and Recommended Standard for Geofoam Applications in Highway Embankments. NCHRP Report 529. Transportation Research Board, Washington, DC, 2004
7. Holtz RD, Christopher BR, Berg RR. Geosynthetic Design and Construction Guidelines, Participant Notebook. NHI Course No. 132013, FHWA HI-95-038 (revised). Federal Highway Administration, Washington, DC, 1998
8. Samtani NC, Nowatzki EA. Soils and Foundations Reference Manual. FHWA-NHI-06-088. National Highway Institute, Federal Highway Administration, Washington, DC, 2006
9. Asphalt Institute. MS-22 Construction of Quality Asphalt Pavements. 3rd ed. Asphalt Institute, Lexington, KY, 2020
10. Asphalt Institute. MS-4 The Asphalt Handbook. 7th ed. Asphalt Institute, Lexington, KY

Chapter 7
Pavement Maintenance and Management

7.1 Pavement Management System

7.1.1 Definition of Pavement Management System (PMS)

Definition of PMS

The road surface management system involves the application of a system analysis method; comprehensive consideration of technical, economic, social, and political factors; coordination of road surface management activities; and promotion of road surface management proccssatization. It provides analytical tools and methods for the decision makers of the management department to help them consider, analyze, and compare the possible countermeasures; to quantitatively estimate the after-effects of the countermeasures; and to select the best cost-effective plan on the basis of predetermined standards and constraints. Therefore, the establishment and implementation of a pavement management system can help the management department improve the effect of decision-making, expand the scope of decision-making, provide feedback information on the effect of decision-making, accumulate management experience, and ensure the coordination and consistency of decision-making units at all levels within the department.

Functions of PMSs

The main functions of the PMS are as follows:
   Storing information about pavement segments;

(A) Identifying funding needs;
(B) Reporting on the state of pavement;
(C) Prioritizing maintenance and rehab funding;

© Tongji University Press Co., Ltd. 2026

H. Li et al., *Road Engineering*, https://doi.org/10.1007/978-981-95-6659-4_7

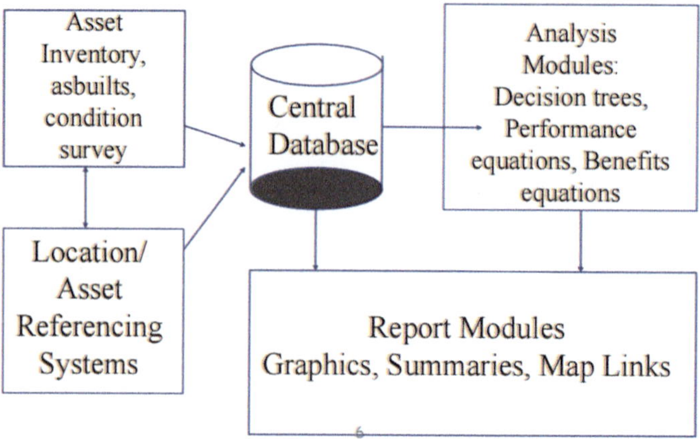

Fig. 7.1 Framework of the PMS

(D) Predicting the effects of M&R activities and funding.

PMS Framework

The first step in PMS work involves collecting pavement information and placing it in a database. Then, the condition of the pavement is evaluated. Next, resources are allocated following policy guidelines. Finally, the performance of individual projects and networks as a whole can be predicted. The framework of the PMS is as follows (Fig. 7.1).

Pavement Management System Objectives

The PMS has the following four main objectives:

(A) Enhancing knowledge of inventory and asset value;
(B) Developing links that tie resource allocations to savings from replacement;
(C) Establishing standardized processes and protocols;
(D) Facilitating life-cycle costing in decision-making.

7.1.2 History of PMS

Systematic pavement management first appeared in Canada in the mid-1960s, initially for pavement design[1, 2], the main purpose of the system is to expand the road network and eliminate road network technical defects such as road surface, alignment, and road capacity. In the 1970s, the demand for road travel surged, but the investment in highway maintenance resources was limited. The demand for optimal resource allocation for road maintenance projects has gradually received attention, and the system has gradually been transferred to the application of pavement maintenance management. The term pavement management system (PMS) first appeared in 1971. Later, many researchers began to work on the establishment of PMSs[3, 4, and by the late 1970s, simple PMSs were being tested. In the 1980s, the international development of PMSs increased, and many states and provinces in the United States, Canada, Europe, Australia, and Japan successively established and implemented PMSs. Owing to the large investment in highways, the benefits of scientific decision-making or the losses caused by decision-making errors are considerable. To promote scientific decision-making, the United States promulgated and implemented the famous Ice Tea Act Intermodal Surface Transportation Efficiency Act (ISTEA) in 1991, and the PMS became one of six regulatory systems required by federal law. The act requires that all roads in the United States that receive federal funds be managed by the PMS and formally requires relevant road management departments to complete the construction of the PMS before October 1995. It stipulates that the construction of a PMS directly affects the amount and priority of the federal auxiliary budget. Therefore, relevant road management departments began to use PMSs to manage their road networks. Since then, PMS has entered a period of gradual application and steady development.

The above period of PMS can be roughly divided into four major stages of development.

The first stage is the initial generation of the PMS, which is essentially a pavement database system with simple system functions and structures. This can be called the first-generation management system; strictly speaking, this stage of the system is not a true sense of the PMS.

In the second stage, on the basis of the pavement database system, the subsystems of pavement service performance evaluation, service performance prediction, and decision are added to form a more perfect PMS. This stage is a very important stage in the development of PMSs, during which the system architecture has become relatively perfect. PMS research was very active during this period, and many forward-looking, groundbreaking ideas were produced during this period. This fact is reflected in the first and second pavement management conferences held in Toronto, Canada, in 1985 and 1987 and in their literature. During this period, the practical application of PMS was also very active, and a series of application systems were developed and established. The development of many systems created a good practice platform for theoretical research, which was also an important factor in the rapid development and fruitful results of PMS research during this period. The system of this period can be called the second-generation system.

However, the system at this stage still has the following shortcomings: first, the principles (knowledge) of pavement evaluation and decision-making and the process of evaluation and decision-making are mixed together, which is not convenient for improving the principles of evaluation and decision-making with the passage of time and is not easy to transplant; second, each link of the system can provide only a decision that is in line with existing policies and habits. With improvements in economic level and technological level, some high-level managers are highly concerned with future pavement investment policies and the rationality of pavement design ideas and standards, but the system cannot provide comparative information in this respect. Third, a road network can have only one standard and lacks hierarchy. Fourth, the system cannot be easily expanded. The reasons for these shortcomings are that the division of labor between man and machine is not clear in the study of the system, the decision of the management personnel is intentionally or unintentionally replaced by the system decision, and the designed system lacks the function of analysis and comparison.

To improve the above shortcomings, from the late 1980s to the middle to late 1990s, PMSs entered the third stage of development. Compared with the system of the second stage, the system of the third stage clearly defines the division of labor between man and machine; that is, the system helps, rather than replaces, the manager to evaluate and analyze the change in road conditions, analyzes the impact of investment policy and technology policy, and the system clearly returns to the position of assistant and adviser. The system function is adjusted, the traditional planning and planning of the main body position are relegated to a secondary position, the function of the data structures in software design is changed, the expert knowledge from system operation in the formation of an independent expert knowledge database is eliminated, the real separation of evaluation and decision-making principles and processes is realized, more comparisons, analyses, and suggestions in terms of function are added, the system function of "decision-making" is transformed into "auxiliary decision-making", and a "technical" system is gradually transformed into a "tool" system. The structure and function of the system are more scientific and reasonable and are more convenient for the promotion and application of the system.

The third-generation system has also made significant progress in data acquisition and achievement expression and presentation. With respect to software, the Windows operating system is widely used to replace the original DOS operating system and has gradually adopted computer network technology. With respect to data collection, a large number of efficient, automated equipment items, and multimedia technologies are widely used to replace the original labor-intensive manual work. In the display of results, geographic information systems (GISs) are widely used as the operating environment of the system and the display platform of analysis results, and multimedia technology is widely used, which greatly enhances the degree of visualization of the system (Table 7.1).

At present, the system is developing into the fourth generation, considering more intelligence, overcoming discrete management, realizing multispace synchronous management, transitioning from traditional single-facility management to multifacility management, asset management, and a more scientific index system transition.

China began to approach PMSs in the mid-1980s. In 1984, the Beijing Highway Department took the lead in developing highway pavement databases in cooperation with the Electronic Technology Research Institute of the Ministry of Petroleum. In 1985, the Highway Research Institute of the Ministry of Communications cooperated with the United Kingdom to carry out a district-level PMS transplantation attempt in Yingkou, Liaoning Province. In 1986, Beijing Road, in cooperation with

Table 7.1 Historical development of PMS in foreign countries and China (1950–2000)

	1950	1960	1970	1980	1990	2000	Events
Abroad							The road network expanded after World War II
							PMS is generated, the first stage
							PMS Development, the second stage
							PMS Development, the third stage
							PMS Development, the fourth stage
China							After the reform and opening up, the road network expanded
							PMS Development, the second stage
							PMS Development, the third stage
							PMS Development, the fourth stage

Tongji University of Road and Traffic Engineering, started working on the research and implementation of the PMS. The same year, Tongji University and other units jointly undertook the Ministry of Communications Highway National Science and Technology Research plan "Trunk Highway Evaluation of Asphalt Pavement Maintenance Technology Research", marking an upsurge in PMS research and practice in China. BJNPMS [system], the first phase of cooperation between the Beijing Highway Department and the Department of Road and Traffic Engineering of Tongji University (BJNPMS system), was completed and certified in 1989; it was the first PMS independently developed and actually put into use in China. Since then, the system has been upgraded many times and has been in continuous operation successfully. Since then, the research and development of pavement and bridge management systems has been a popular research topic in the field of highways in China. According to incomplete statistics, until the 1990s, many scientific research units cooperated with many provincial, city, and regional highway and municipal administration departments and developed more than 20 kinds of PMSs successively, and more than 50 units had PMSs. However, unfortunately, the actual normal operation of the system is not common, the real environment is very unsatisfactory, the application demand is not exuberant, and relatively backward technical practices exist, resulting in the application of PMSs being tested.

Similarly, with respect to the development of PMSs in China in the above historical period, on the whole, it has the characteristics of a high starting point but a thin foundation and rapid development but poor practice. The mid-to-late 1980s was the most prosperous period of PMS research in the world, during which the functions and frameworks of PMSs were essentially perfected, and it was in a period of transition from the second stage to the third stage. At this time, domestic scientific research was in the recovery stage after a long period of silence, and the increase in foreign research naturally led to the response of domestic researchers. At the same time, domestic infrastructure is increasing. Many road projects financed by the World Bank or ADB clearly require the establishment of PMSs, which provides certain opportunities and possibilities for domestic PMS research. Therefore, domestic research and practice of PMSs began during this period. However, owing to the weak theoretical research foundation, most departments could only take the path of combining technology introduction and independent research and development at the beginning, which also made it unlikely that China experienced the first stage of PMS development but could have directly entered the second stage at the beginning. A small number of domestic researchers have since closely followed the development trends of foreign countries and narrowed the gap between them. It was not until the mid-1990s that they basically caught up with the pace of foreign research in the third stage of PMS development. In practice, the situation at home is far from comparable with that abroad. The blindness and compulsion of the system at the beginning of the establishment, the serious lack of practical application requirements, and the poor use environment led to many systems being either not built before being given up or just built before being shelved. The unsuccessful practice also restricted the development of theoretical research. The domestic research boom soon retreated, and the relevant research entered a relatively calm and slow forward period.

7.2 Pavement Condition Survey and Evaluation

7.2.1 Indices of Pavement Conditions

The main indices used within typical pavement management systems are as follows:

(A) Ride quality
(B) Structural condition
(C) Surface conditions like cracking, patching, potholes, etc.
(D) Safety

Moreover, the requirements for road condition data collection for network-level PMSs and project-level PMSs differ:

(A) Network: collecting information needed to

    a. Identify the overall condition of the network
    b. Identify sections that need work, and
    c. whether structural or functional work is needed

(B) Project: collecting the information required for

    a. Identifying the mechanisms of failure
    b. Updating network information and providing more details
    c. Designing the pavement

7.2.2 Two Main Types of Distress

Asphalt Pavement:

(A) Distresses in the wheel paths

    a. Structural damage caused by heavy loads
    b. Need for structural capacity

(B) Distresses out of the wheel paths

    a. Owing to aging and surface damage
    b. Need for surface restoration

Concrete pavement:

(A) Roughness
(B) Structural damage

7.2.3 International Roughness Index (IRI)

Although the reaction flatness measurement system was a flatness measurement method widely used in the 1970s, because the measurement value of the reaction instrument was high on poor road sections and low on good road sections, so a standard index and method were needed to calibrate it. The National Highway Cooperative Research Program (NCHRP) proposed this problem in Projects 1–18 in 1978, and the concept of the "International Roughness Index" (IRI) was put forward in the subsequent research project "Calibration and Relationship of Reaction flatness Systems". The international flatness experiment conducted by the World Bank in Brazil in 1982 completely and systematically proposed the calculation model and method of the IRI. The IRI was calculated from a one-way longitudinal section; the quarter-car model (Fig. 7.2) was used to drive the known section at a speed of 80 km/h, and the cumulative vertical displacement of the suspension system within a certain driving distance was calculated as the IRI [5].

The IRI is an evaluation index obtained by integrating the advantages of sectional and dynamic flatness measurement methods. It is a dynamic variable obtained from static sectional elevation data after mechanical model calculation. The IRI has the following characteristics: the IRI is related to the dynamic response of vehicle vibration, and the correlation with vehicle performance is established through the 1/4-vehicle model. The IRI is directly related to the elevation of the section, which ensures the time stability of the results. The IRI can be measured by the most widely used instruments (such as a level), and the results are valid; IRI can be converted worldwide (with standard computing procedures) and is portable. Owing to the above characteristics, the IRI has become the most widely used flatness index worldwide (Fig. 7.3).

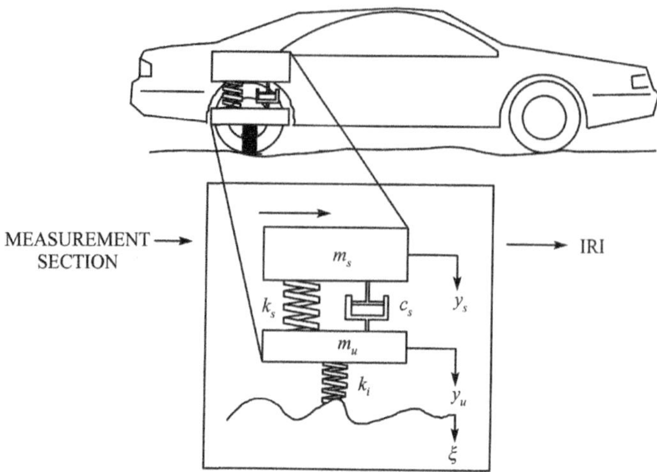

Fig. 7.2 Ideal car model

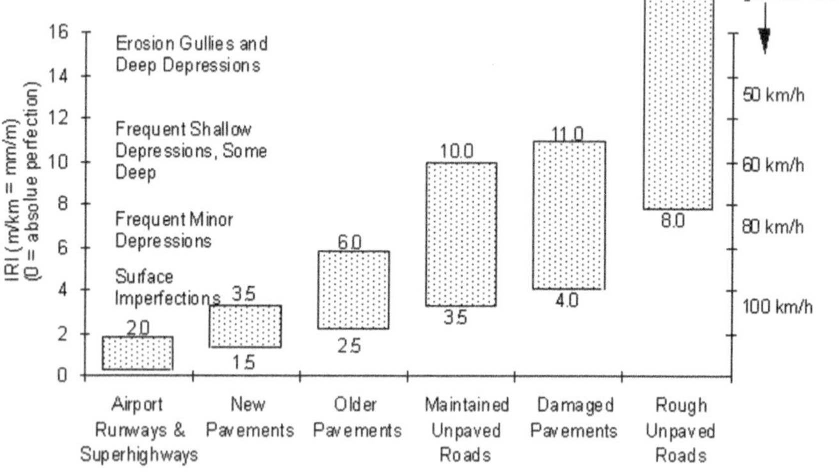

Fig. 7.3 IRI for all types of pavements

7.2.4 *Falling Weight Deflectometer (FWD)*

A falling weight deflectometer applies a pulsed load to a road surface by dropping a specific weight from a specific height. As shown in Fig. 7.4, the applied load can be controlled by the weight and fall of the object, and the duration of the pulse load can be controlled by a buffer such as a rubber pad. Generally, FWD has a load capacity of 15–125 kN, and the equipment used for airfield pavement subsidence measurement is up to 250 kN. The load pulse time is generally between 0.025 and 0.030 s.

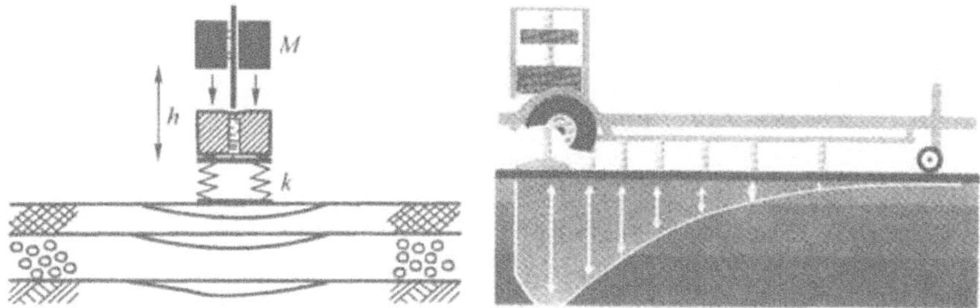

Fig. 7.4 Basic test principles of FWD

The drop weight bending and settling instrument can measure not only the maximum bending and settling under the action of the pavement load but also the bending and settling basin of the pavement, which provides more information for the evaluation of the pavement structure.

When FWD is used to determine the bending and settling of the pavement structure, the position of the sensor should be arranged so that it is suitable for the intersection point between the stress cone and the interface of the pavement structure layers (Yao, 1993), as shown in Fig. 7.4. The load acting on the road surface diffuses to the interior of the road structure. The stress diffusion range forms a stress cone. The intersection point of the stress cone and the interface of each level of the pavement structure has a specific meaning: the bending of the road surface beyond the intersection point is more affected by the characteristics of the structural layers below the intersection point. Therefore, the position of the sensor should be arranged in the bending and settling measurement so that the bending and settling basin test results can reflect the characteristics of each structural layer. Generally, 5–7 sensors are needed for the determination of road bending and settling, and 9 sensors are needed for the determination of airport pavement.

7.2.5 *Traffic Speed Deflectometer (TSD)*

The traffic speed deflectometer (TSD) is currently the world's most advanced bending and settling detection equipment, with the advantages of fast speed measurement and high accuracy, and its detection result is a bending sink. The high-speed laser bending instrument used for road surface bending and sinking data collection does not affect normal traffic, can provide a large amount of scientific research data and data analysis at normal speeds, and can provide test data and analysis support for the bending and sinking value of the road surface to achieve rapid and efficient testing of road surface bending, greatly improving work efficiency and reducing the cost of the work. On the other hand, basic data and analysis support for pavement structures can be provided, which reserves sufficient time for pavement maintenance work and repair work. In addition, it can also support the collection of highway informatization data through rapid and large-scale data collection and then complete the road information database together with other pavement indicators [6].

The working principle of the TSD is based on the Doppler effect. The Austrian physicist Schen Johann Doppler proposed the famous theory of the Doppler effect in 1842, with the following formula:

$$F_d = -F_s \frac{V}{C} \tag{7.1}$$

where

V is the relative velocity of the wave source and receiver.

C is the wave propagation velocity.

F_d is the frequency change of the receiver.

F_s is the transmitting frequency.

Applied to the high-speed laser bending sinker, C is the laser speed of the Doppler laser sensor (speed of light 3.0×10^8 m/s), and F is the reciprocal of time, which can be calculated by the time difference between the laser sensor emission and acceptance, from which the relative velocity V, that is, the bending and sinking speed, can be obtained according to formula (7.1). The high-speed laser bending instrument uses Doppler laser sensors to measure the road deformation speed caused by the high-speed driving of standard trucks and inverts the road surface deformation speed with the road deformation speed. The difference between the sensor's measurements in the sink basin and those of the reference sensor reflects the true motion of the pavement. In the high-speed bending and sinking measurement pavement model, since the bending and sinking speed is affected by the driving speed, to eliminate this influence, the concept of slope is introduced, and the slope relationship is as follows:

$$d\prime(x) = \frac{V_D}{V_K} \tag{7.2}$$

where

$d'(x)$ is the slope at point x of the bending basin.

V_D represents the instantaneous bending and settling velocity.

V_K is the instantaneous driving speed.

7.3 Maintenance and Rehabilitation

7.3.1 Definition

Maintenance of the facilities starts from the first day of opening to traffic, and different maintenance measures are required for different technical conditions of the road. With the passage of time, factors such as oxidation, light aging of pavement materials, acid rain, atmospheric conditions and inappropriate maintenance measures resulting in alkali aggregate reactions of pavement materials, gap and crack damage due to road surface water seepage, and driving comfort and anti-sliding ability will also be constantly attenuated over time. Failure to address these phenomena in a timely manner will have a great impact on the service performance and service life of the structure; to carry out timely maintenance and rehabilitation, it is very important to choose appropriate measures.

Structural damage occurs earlier, often not only because of design, materials, construction, and operation management but also because of other factors. In terms of design, the original structural design of the facility is not reasonable, and unreasonable material design standards, overestimating the strength of the foundation and underestimating the size of the traffic load, etc., are the reasons for early structural damage. In terms of material production, especially for the production of asphalt mixtures, there are often a series of problems, such as the following problems: mixture design standards are not applicable to the size of the traffic volume, the raw material quality is poor, there is no good quality control system in the production process, there is no protection after production, etc. In the construction aspect, the construction standard is too low, the construction personnel's sense of responsibility is poor, the construction method is improper, the site preparation is insufficient, the site strain capacity is insufficient, and the problems found cannot be corrected in time, etc., which can damage the structure earlier. In terms of operational management, the factors include traffic volume that exceeds expectations, overloaded vehicles, drainage system failure, and unauthorized temporary use of facilities. The factors of maintenance include the poor engineering quality of the maintenance measures themselves, the maintenance not being carried out in a timely manner, the failure to compensate for a number of problems existing in the construction in time, and the failure to find the initial symptoms of damage in the structure. Owing to the existence of many factors and many links, coupled with field operations, the construction process has difficulty realizing industrialization and is affected by weather; thus, the quality of a project is difficult to guarantee inevitably, and different degrees of problems are difficult to avoid. It is necessary and routine to compensate by maintenance and repair.

At present, two strategies are commonly used for maintenance: the "damage is repair" type of maintenance, also known as maintenance, which refers to the maintenance of the structure with obvious damage that affects its normal use; and preventive maintenance, which is adopted to prolong the service life and protect the original structure when the structure has not been obviously damaged and the service function is still sufficient to meet the requirements. Because the meaning of "prevention" is difficult to define, it is more commonly called "protective conservation" worldwide.

Broadly speaking, maintenance includes maintenance, which is a maintenance measure; in a narrow sense, maintenance and repair are two different technical measures. In Japan, maintenance refers to the act of maintaining the function of a road, whereas maintenance refers to the act of enhancing the function of a road. In Europe, the United States, and in our own country, it has a similar meaning.

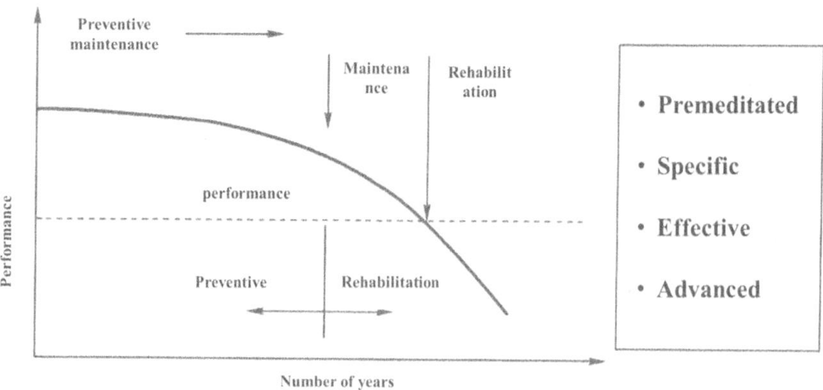

Fig. 7.5 Maintenance of roads at different stages

Figure 7.5 shows the applicable maintenance technical measures when the technical condition of the facility is at different stages. It is very effective to take holistic measures to extend the life of the facility before performing intermediate repairs. These measures can be called preventive or protective conservation measures, and the corresponding concepts become preventive or protective conservation concepts. Regardless of the type of measure, conservation work should be planned and be reflected in the initiative to conserve rather than in passive coping; the measures taken should be targeted and effective. Only by focusing on specific problems can we achieve good results and save maintenance funds. The timing of maintenance should be in advance, which is also one of the main means to ensure the effect and reduce the cost.

7.3.2   *Flexible Pavement Maintenance and Rehabilitation Strategies*

At present, commonly used asphalt pavement maintenance techniques are shown in Table 7.2. According to the scope of these measures, the technologies in the table can be divided into two categories, namely, local conservation technologies and holistic conservation technologies.

(1)   **Crack Sealing, Pit Repair and Excavation**

These three measures are very widely used daily maintenance measures.

Crack sealing is used to seal cracks to reduce moisture infiltration into the pavement. The commonly used crack sealing materials include ordinary asphalt, modified asphalt, special sealing materials, and sealants, etc. The crack effect of ordinary asphalt sealing is poor, its use is short, it is easy to remove vehicle tires in the summer, and it is easy to pollute the road surface; additionally, it cannot play the role of filling,

Table 7.2 Commonly used asphalt pavement maintenance techniques

Serial number	Maintenance measures	Scope of application
1	Crack sealing, pit repair, excavation	Local
2	Repair ruts	
3	The fog sealing layer	Global
4	Asphalt regeneration	
5	Thin slurry seal and micro surface	
6	Seal and thin cover	
7	The surface	
8	Road regeneration	
9	Renovate, reinforce or rebuild	

which is generally used when the traffic volume is very small for maintenance of the road surface. Modified asphalt is much more effective and is suitable for roads with high traffic volume. In fact, the special sealing material is also a modified asphalt product in which filler is added, which is suitable for the maintenance of heavy traffic road surfaces. Rubber-modified asphalt is generally used as a sealing material, and the dosage of the modifier is relatively high. This kind of modified asphalt has high elasticity and strong denaturation ability and is more suitable for the deformation of cracks. The filling of cracks generally involves three processes: cutting or expanding seams, clearing seams, and irrigating seams.

Pothole repair refers to the repair of potholes on a road surface. Under normal circumstances, the appearance of potholes on a road surface reflects road dystrophy. Potholes are often found in China because of the poor uniformity of the material, especially on highways after rain. When performing pit repair, one side of the patch should be parallel to the centerline. When the pit area is large, the material used should be as consistent as possible with the original pavement structure.

Digging and filling involves excavating pieces of damaged pavement that have not yet formed pits and repairing and maintaining the original pavement structure materials. The more serious local damage to the road surface or the local segregation of the road surface will cause severe road seepage; if mearures not taken, it will have a serious impact on the whole road structure. At this time, even if the surface of the road has not yet formed a pit or is relatively complete, it is necessary to take repair maintenance measures to protect the entire pavement structure and prolong its service life.

(2) **Repair Ruts**

The appearance of ruts affects road safety and ride comfort. Generally, there are three measures for the maintenance of road ruts, namely, milling, filling, and wheel-track milling.

When the ruts are light and the asphalt pavement is thick, the raised part of the ruts can be milled out to mitigate the effect of the ruts. The second measure involves laying a layer of material on the depressed part of the rut (such as the microsurface) or milling along the wheel track to a certain depth and laying a new asphalt mixture.

(3) Fog Sealing Layer

After a certain number of years, the surface of the asphalt will experience a certain degree of aging. The effect of the fog sealing layer is to restore this aging, and it has a certain sealing effect on the microcracks. It is suitable for the maintenance of pavement with sufficient strength and a smooth and intact pavement to restore the aging of asphalt and prolong the life of the pavement.

(4) Asphalt Regeneration

The fog sealing layer mainly restores the performance of the asphalt on the pavement surface and has a certain sealing effect on microcracks. For pavements with cracks of a certain depth or some bottom-up cracks or larger voids, more severe road surface seepage can occur, and a fog sealing layer has difficulty obtaining a good maintenance effect. However, other types of pavement maintenance technologies, namely, asphalt recovery or asphalt regeneration technologies, can yield better results. These technologies takes advantage of some characteristics of emulsified or diluted asphalt to produce a highly permeable asphalt liquid that penetrates the pavement structure to restore asphalt properties, seal cracks and voids, reduce aggregate looseness, and reduce pavement permeability, thus prolonging the life of the pavement.

The applicable conditions of asphalt recovery or asphalt regeneration technologies are similar to those of fog sealing; that is, the pavement conditions are good, the pavement is smooth, and the pavement strength meets the use requirements. The damage to the road surface is mainly caused by aging, loosening, or the presence of more microcracks.

(5) Thin Slurry Seal and Microsurface

A diluted slurry seal involves mixing emulsified asphalt, aggregates with specific gradations, filler, water, and additives; mixing them into a dilute slurry mixture according to a specific proportion under normal temperature conditions and spreading it on the road to form a 3–12 mm thick seal. Its main function is to seal the road surface with a 3–12 mm thick sealing layer, cover small cracks, delay the loosening and aging of the road surface, restore and improve the skid resistance and driving quality of the road surface, and improve the appearance of the road surface.

The microsurface is the replacement product of the dilute slurry seal; the binder used is polymer-modified emulsified asphalt, and the other components are the same as those of the dilute slurry seal. The thickness of the single-layer microsurface is generally <10 mm. The thickness of the two-layer microsurface construction is generally approximately 15 mm.

(6) Seal and Thin Cover

The main function of the sealing layer is to seal the road surface, cover small cracks on the road surface, prevent moisture from entering, delay the aging and loosening of the road surface, and restore or improve the skid resistance of the road surface.

There are many forms of pavement sealing, and gravel sealing is currently commonly used. The gravel seal can be constructed as a single or double layer. A single layer of gravel seal is first sprinkled on the road surface with a layer of asphalt (diluted asphalt or emulsified asphalt), then a layer of aggregate is spread, and the tire roller compaction is immediately applied, forming a seal. The double layer of crushed stone seal is constructed in two layers. The first layer consists of aggregates with larger particle sizes, whose quantity accounts for 50–70% of the total aggregate quantity. In the second layer, when other materials with smaller particle sizes are used, the maximum particle size of the second layer is generally half of the maximum particle size of the first layer.

A thin cover refers to a cover below 4 cm, which can not only improve the skid resistance, delay aging, reduce water seepage, and perform other functions of the road surface but also significantly improve the driving quality of the road surface and is suitable for moderate cracking (longitudinal, transverse, and cracking), moderate looseness, and wear. It is difficult to achieve thin coatings with ordinary hot mix asphalt mixture technology, and warm mix technology may be a good choice.

(7) **The Surface**

Cladding generally refers to the use of a hot mix asphalt mixture to cover the old asphalt pavement to a thickness of 4 cm or more, which is suitable for the maintenance of the pavement in medium to poor condition. The design requirements of hot mix asphalt mixtures are related to the traffic volume, and the requirements of new construction are the same. The thickness of the cover is generally determined empirically, depending on factors such as the condition of the old road surface, traffic volume, and local climate conditions.

(8) **Road Regeneration**

Pavement regeneration is an increasingly common maintenance technique. From the point of view of regeneration technology, road regeneration can be divided into cold regeneration technology and hot regeneration technology. The construction methods can be divided into in situ (site) regeneration and factory (plant) regeneration. By combining these two classification methods, pavement regeneration can be divided into four types: in-place cold regeneration, in-place hot regeneration, plant cold regeneration, and plant hot regeneration. The asphalt material of the original pavement structure can be used as the surface layer, base, or bottom base of the new pavement structure after regeneration. The above four types of regeneration technologies are suitable for situations where the pavement infrastructure is intact, the bearing capacity can meet the traffic requirements, and the pavement damage is mainly aging and cracking. Some severely damaged road surfaces, which involve the foundation of the road surface, can use the reuse technology of road material, that is, the recycling of the base, also known as the full-depth of recycling technology.

(9) **Renovating, Reinforcing, or Rebuilding**

Complete transformation of the pavement structure or reconstruction of the pavement structure combined with rerouting, expansion, etc.

7.3.3 Rigid Pavement Maintenance and Rehabilitation Strategies

The common techniques of cement concrete pavement maintenance include full-thickness repair, partial-thickness repair, joint repair, diamond grinding, road shoulder repair, drainage system repair, joints, and crack sealing. Each technique is designed to prevent and repair some damage or a combination of damages. While each technique can be used in isolation, it can be more effective when several measures are used.

Pavement maintenance can be divided into preventive maintenance and maintenance. The latest definition of preventive maintenance by the AASHTO Road Standing Committee is as follows: preventive maintenance is a planned and cost–benefit-based maintenance strategy for the existing road system when the road surface is in good condition. Preventive maintenance can delay damage to the road surface, maintain or improve the existing traffic conditions of the road surface, and postpone expensive overhaul and reconstruction activities by prolonging the service life of the original road surface without improving the structural capacity of the road surface.

(1) Grouting Technology for Cement Pavement

An empty bottom plate is an important cause of damage to cement pavement. If the clearance is not identified and treated in a timely manner, the damage is bound to speed up, and the road surface has difficulty reaching its expected life. The general method of treating plate bottom empties involves the use of grouting technology to seal the plate bottom. Because of the high requirements for construction, grouting is not so much a technology as a process. If used incorrectly, it may accelerate the fracture and fragmentation of plates.

(2) Cement Pavement Plate Replacement/full Thickness Repair

The replacement of a full-thickness concrete slab restores the ability to transfer loads between lateral joints or cracks and minimizes vertical bending. The type and severity of cement pavement damage are investigated to determine whether the plate should be replaced. The identification and quantitative evaluation of pavement damage through preliminary road condition investigations is an important step in the development of a suitable repair plan. In the case of severe damage over a large area, the whole plate should be removed and replaced. When the damage area is small, identification is performed to determine the boundary where full-thickness repair is needed.

Properly designed and constructed full-thickness concrete repairs can maintain satisfactory performance for a period of 10 years or more. According to foreign studies, load transfer systems should be considered when repairing the full thickness of most heavy vehicle routes. A load-transfer system may not be required for pavements without dowel bars if they are constructed on a stable treated base or are located in dry climatic conditions.

(3) Joint/Caulk Repair Technology

Under the action of the environment and load, the performance of cement pavement gradually decreases. Among the environmental factors, the impact of water on pavement damage is the greatest. To avoid or reduce the infiltration of surface water into the pavement structure, timely and effective sealing and replacement of various joints and cracks on the pavement are necessary. In addition, joint/crack sealing can prevent hard foreign bodies from entering the joints/cracks, which hinders the expansion and deformation of the plate due to heat and causes damage, such as spalling and disintegration of the seam edge. In addition to the above two functions, joint sealing can also reduce the corrosion of the force transfer rod and prolong its service life. The joint walls should be in good condition with little or no peeling. When the road surface is relatively new, neoprene rubber is used for sealing; when the pavement is used for a long time (more than 10 years), silicone is often used to seal it.

(4) Joint/Caulk Load Transfer Capacity Recovery Technology

Dowel Bar Retrofit (Dowel Bar Retrofit) is used to repair lateral joints and cracks for better load transfer capability. With increasing service life, when the interlocking effect of joints or cracks is lost, it is necessary to consider the use of dowel bar retrofit measures.

(5) Diamond Grinding Technology

The temporary improvement in the skid resistance enhances the skid resistance of the tire and improves safety. The use of microgrooves on the road surface could also improve safety. These tiny grooves allow for improved drainage between the tire and the road, which increases the tire's skid resistance.

(6) Cement Pavement Fragmentation and Petrochemical Technology

The cement pavement can be pulverized to break the cement pavement plate into relatively small pieces. In this way, the characteristics and problems existing on the old cement road surface, such as joints, cracks, and faults, will not reappear in the HMA overlay. This can prolong the service life of the overlay and ensure the flatness of the road surface. High efficiency, which is more economical than other repair measures, reduces user inconvenience. Notably, for well-structured pavements and road surfaces, rubblization should not be performed. It is not advisable to break up the pavement simply to reduce reflective cracks.

7.4 Life Cycle Cost Analysis and Life Cycle Assessment

Life cycle cost analysis (LCCA) and life cycle assessment (LCA), as methods that extensively cover the whole life cycle of the evaluation and analysis object, can carry out effective and accurate economic and environmental assessments of pavement and provide support for promoting high-quality and low-carbon transformation in pavement engineering. This chapter explains life cycle cost analysis (LCCA) and life cycle assessment (LCA) in detail around life cycle analysis methods and further demonstrates the application of the two methods through evaluation cases. In addition, product category rules (PCRs) and environmental product claims (EPDs) are introduced.

In the past, we paid more attention to the initial cost of highway construction and did not pay enough attention to the later costs of operation and maintenance, especially the long-term impact of highway construction on the environment. However, a highway project is not a one-time consumer good; after it is completed and put into operation, it also experiences many maintenance cycles. To maintain its normal use function, the operation, maintenance, and maintenance costs cannot be underestimated. For a variety of different road design schemes, it is difficult for us to intuitively judge the advantages and disadvantages of various schemes, and we must use a scientific analysis tool—life cycle cost analysis—to carry out economic comparisons. The whole life cycle cost analysis plays an important role in pavement design, and it is an essential tool for optimizing pavement design schemes.

Climate change is a major global challenge facing mankind today. As the largest emitter, China plays a crucial role in the fight against climate change. Following the 2015 Paris Climate Change Conference, most governments, including the European Union, the United States, China, and India, agreed to abide by the Paris Agreement to limit global warming to 2 °C. As one of the three major areas of carbon emissions, the transportation sector has a long way to go to save energy and reduce emissions.

7.4.1 Life Cycle Cost Analysis

Life cycle cost analysis is a technique for evaluating the long-term economic efficiency of alternative investment options on the basis of economic analysis principles. It takes into account the initial construction costs of alternative investment options as well as future administrative costs, user costs, and other related costs during the analysis period. Its purpose is to determine the best value of investment benefit, that is, to obtain the solution with the lowest long-term cost under the desired performance objective.

As early as 1960, the AASHTO introduced the concept of life cycle cost analysis. Additionally, in the 1960s, the United States established two programs to promote the application of the life-cycle cost principle in pavement design and pavement type selection: the National Cooperative Highway Research Program (NCHRP) and a project funded by the Texas Department of Highways and Transportation to develop a rigid pavement system that requires life-cycle cost analysis to rate alternatives. These two projects provide a research basis for life cycle cost analysis in the pavement area. Both the 1986 and 1993 AASHTO road design guidelines actively advocate life cycle cost analysis and discuss in detail what costs should be included in life cycle cost analysis. Life-cycle cost analysis has also been developed in other countries, such as Canada, Australia, and Egypt. To date, the method has been constantly supplemented and developed, and the cost factors are increasingly comprehensive.

There are many methods for analyzing life-cycle costs, such as the present value method, the annual cost method, the rate of return method, and the benefit–cost ratio method. This study adopts the present value method for analysis; that is, the costs and benefits occurring at different times in the analysis period are converted into current costs and benefits according to a predetermined discount rate to facilitate comparisons on a common basis. The present value method includes cost present value and net present value. Since the main purpose of cost analysis and comparison is not to make accurate cost or benefit estimations for each design scheme but rather to make a relative evaluation of the economic value of each design scheme, it is not necessary to consider all costs or benefits completely and accurately. Instead, only the main costs and benefits and related main parameters that affect the evaluation results of each program should be considered, and the selection of each program should be coordinated [5]. Considering that some of the benefits associated with road performance are difficult to measure, only costs are involved here without considering benefits; that is, the cost present value method is adopted. To meet the needs of different user levels, two different economic optimization indices are considered, namely, total engineering cost (excluding the user fee) and total cost (including the user fee). The calculation formula is as follows:

$$ZZJ_{x1,n} = IC_{x1} + \sum pwf_{i,yi}RC_{x1,y1} + \sum pwf_{i,t}MC_{x1,t} - pwf_{i,n}SV_{x1,n}$$

$$ZFY_{x1,n} = IC_{x1} + \sum pwf_{i,yi}RC_{x1,y1} + \sum pwf_{i,t}(MC_{x1,t} + UC_{x1,t}) - pwf_{i,n}SV_{x1,n}$$

where

$ZZJ_{x1,n}$—The total cost of plan $x1$ in n years.

$ZFY_{x1,n}$—The total expense of plan $x1$ over n years.

IC_{x1}—Initial construction cost of plan $x1$.

$RC_{x1,y1}$—Conversion costs at the end of different phases of scheme $x1$.

$MC_{x1,t}$, $UC_{x1,t}$—The maintenance cost and user cost of scheme $x1$ in year t.

$SV_{x1,n}$—Residual value of plan $x1$.

$pwf_{i,yi}$, $pwf_{i,t}$, $pwf_{i,n}$—The discount factor.

On the basis of the LCCA, the World Bank developed a fully functional HDM-4 model in 2003. As with professional road economic analysis software, it can consider external factors (weather, road maintenance measures, etc.) and internal factors (traffic changes, etc.) through technical and economic evaluation from angles such as the user cost and engineering cost to determine the optimal road maintenance strategy and assess the investment value. In HDM-4, each analyzed road surface is defined by a set of unique parameters, such as geometric characteristics, environmental conditions, traffic load, road surface conditions, and road surface design. As long as the initial road surface design and maintenance measures of a certain road surface are given, the present value of the corresponding life cycle can be calculated during the analysis period. HDM-4 selects the best option for different strategic schemes, and the formulation of planning or maintenance strategic schemes itself is not one of the functions of HDM-4. In essence, it is road economic analysis and evaluation software. The core of its evaluation is cost, which is not only the narrow sense of infrastructure planning or maintenance cost (engineering cost) but also the total social cost, including user cost. The more accurate and comprehensive user cost calculation is the advantage of HDM-4.

7.4.2 Life Cycle Assessment

Life cycle environmental assessment is a process of evaluating the environmental load associated with the whole life cycle stages of a product, process, or activity, from raw material collection to production, transportation, sale, use, recycling, maintenance, and final disposal of the product. It first identifies and quantifies energy and material consumption and environmental release throughout the whole life cycle stages, then evaluates the impact of consumption and release on the environment, and finally, it identifies and evaluates the opportunities to reduce these impacts [7].

As an environmental management tool, it can not only carry out effective quantitative analysis and evaluation of current environmental conflicts but also evaluate the environmental problems involved in the whole process of products and their "cradle-to-grave" life cycle; therefore, it is an important supporting tool for product environmental management. It can not only be used for enterprise product development and design but also effectively support the government's environmental management departments in formulating relevant environmental policies. At the same time, it can also provide clear product environmental labels to guide consumers' consumption behavior toward environmental products [8].

In 1997, the ISO14040 standard divided LCA implementation steps into four parts: objective and scope definition, inventory analysis, impact assessment, and result interpretation [9].

(1) Definition of Objectives and Scope

The definition of objectives and scope is the first step in life cycle assessment, which is the starting point and foothold on which inventory analysis, impact assessment, and result interpretation depend; it determines the subsequent stages and LCA evaluation results and directly affects the whole evaluation process and the final research conclusion [8].

The objective definition should clearly state the purpose and reasons for undertaking the life cycle environmental assessment and the intended application areas of the findings. Scoping needs to consider and clearly describe the functions of the product system, functional units, system boundaries, data allocation procedures, types of environmental impacts, data requirements, assumptions, constraints, original data quality requirements, types of review of results, and the types and forms of reports required for research. The definition of the research scope should be sufficient to ensure that the breadth, depth, and detail of the research are consistent with the required goals so that all the processes of the life cycle of the studied object fall within the boundary of the system.

1. Function, function unit, and reference flow

There are many functions in a system, and the selection of functions in a study depends on the goal and scope of the life cycle assessment. The functional unit defines the quantification of theperformance characteristics of the product. The main purpose of functional units is to provide input- and output-related references. These references play a crucial role in ensuring the comparability of life cycle environmental assessment results. The comparability of life cycle assessment results is particularly important when different systems are evaluated to ensure that relevant comparisons are made on a common basis. To achieve the desired functionality, it is important to determine the reference flow in each product system, that is, the number of products required to achieve the functionality.

When the target of the research is a pavement system, the determination of functional units is questionable. Owing to the diverse characteristics and functions of pavement systems, realizing the unification of functional units when comparing two different types, processes, materials, or pavement structures is both difficult and unrealistic. The function of the road itself includes not only the bearing capacity of the traffic but also the bearing capacity and deformation resistance of the structure. In addition to the service life of the road, environmental toughness, etc., these characteristics in different types, processes, materials, or structural pavement comparisons cannot be completely consistent. The general practice of existing methods is to maintain the consistency of the main functions as much as possible and to account for other inconsistent functions. The general definition includes the traffic volume of the road, expressed by the annual average daily traffic volume or equivalent standard axle load times and the traffic growth rate; it also includes the length of the analysis, which is typically 10–50 years. There are also factors such as special climatic and geological conditions that roads may face.

2. System boundary

Life cycle environmental assessment is carried out by defining the product system as a model describing the key elements of the physical system. The system boundary defines the unit processes to be included in the system. Ideally, the inputs and outputs at the boundaries of the production system should be basic flows. When conditions permit, a complete life cycle environmental assessment should include all stages of the life cycle. These factors and stages can be ignored only when they have little influence on the final conclusion, and the ignored conditions are also called boundary conditions. When certain influential factors are omitted, the reasons and possible effects must be stated. In many cases, the system boundaries initially defined need to be subsequently refined [6].

3. Data quality requirements

Data quality requirements generally specify the characteristics of the data required for research. The description of data quality is important for understanding the reliability of the findings and for correctly interpreting the findings.

(2) **Checklist Analysis**

Life cycle inventory (LCI) analysis is a data-based and objective quantification process of energy consumption, material consumption, and environmental emissions (including exhaust gas, wastewater, solid waste, and other environmental releases) of a product, process, and activity during their life cycle. Inventory analysis quantifies the relevant inputs and outputs of a product system through data collection and computation procedures. As the core part of LCA and the basis of environmental impact assessment, this analysis and evaluation run through the whole life cycle of products and constitute the most well-developed part of LCA at present.

Owing to the complexity and long-term nature of the pavement system, the inventory analysis process is generally divided into several stages. The common method divides the whole life of the pavement into five stages: the raw material acquisition stage, the construction stage, the use stage, the maintenance stage, and the recovery stage.

(3) **Impact Assessment**

Life cycle impact assessment (LCIA) is an important part of LCA. It involves the classification and calculation of data and aims to characterize and evaluate the environmental impact factors identified by inventory analysis qualitatively and quantitatively to determine the degree of impact of material and energy exchange on the external environment in the process of the product life cycle.

According to the technical framework formulated by ISO14042 (2000), the life cycle impact assessment is divided into two parts: essential elements and optional elements. The essential elements are used to transform the inventory analysis results into parameter results. The optional elements are used to reduce the parameter results to a single comparable index and verify the data quality to determine the real reliability of the obtained evaluation results.

(4) **Interpretation of Results**

Interpretation of the results refers to summarizing the results of the checklist analysis and impact evaluation to form conclusions and suggestions. In LCA, if the investigation scope, the definition and distribution method of the system boundary in inventory analysis, and the selection of characteristic coefficients in the impact evaluation stage differ, the conclusions may differ. Therefore, it is necessary to interpret the results. Moreover, most of the data used in inventory analysis are not actually measured and usually contain speculative and quoted data. In international standards, it is necessary to evaluate data quality, such as the completeness and representativeness of the data used. There have been sensitivity analysis, uncertainty analysis, and research on data quality for a long time, but the development of this link in the field of LCA research is still relatively slow.

Discussion, Questions and Exercises

(1) What is a pavement management system, and what are the main functions of PMSs?

(2) Describe the specific steps in the PMS framework for gathering pavement information and assessing pavement conditions, and explain why these steps are critical for resource allocation and project performance prediction.

(3) Discuss how TSD equipment uses the Doppler effect to measure pavement bending and settling, and explain how it can improve work efficiency and reduce costs.

(4) In pavement maintenance and rehabilitation, explain how preventive maintenance is different from remedial maintenance.

References

1. Hudson, W.R., Finn, F.N., McCullough, B.F., Nair, K. and Vallerga, B.A., 1968. Systems Approach to Pavement Design, Systems Formulation, Performance Definition and Materials Characterization. Final Report, NCHRP Project, 1(10). https://onlinepubs.trb.org/Onlinepubs/nchrp/nchrp_rpt_139.pdf
2. Scrivner, F.H., Moore, W.M., McFarland, W.F. and Carey, G.R., 1968. A systems approach to the flexible pavement design problem. https://rosap.ntl.bts.gov/view/dot/85089
3. Pavement Management Guide Roads and Transportation Association of Canada. Pavement Management Committee RTAC, 1977 the University of California
4. Haas, R. C. G., & Hudson, W. R. (1978). Pavement management systems. McGraw-Hill. https://trid.trb.org/View/77151
5. SUN L J. Road and Airport Facility Management[M]. Beijing: China Communications Press, 2009
6. GONG J, CHANG C L, ZHANG Z Y, et al. Calibration Method of High-speed Laser Bending Instrument[J]. Journal of Beijing University of Technology, 2015, 41(08): 1200-1205
7. YAO Z K. Pavement Management System[M]. Beijing: China Communications Press, 1993
8. ZHENG X J, HU B. Literature Review of Life Cycle Assessment (LCA) in China and the Latest Research Progress Abroad [J]. Scientific and technological progress and Countermeasures. 2013
9. ORGANIZATION I S. ISO 14040: Environmental Management-Life Cycle Assessment-Principles and Framework [M]. 1997

Chapter 8
Sustainable Pavement

With the rapid development of urbanization and the continuous expansion of the scale of transportation infrastructure, urban roads, highways, airports and port pavements, squares, parking lots, and other road pavement facilities are large and wide, and the total urban paving rate exceeds 20%. However, traditional large-scale impermeable asphalt pavement, while satisfying traffic functions, introduces many hazards to the environment, significantly changes the surrounding environment and ecological system, and causes urban flooding, nonpoint source (NPS) pollution, the urban heat island (UHI) effect, traffic noise, greenhouse gas emissions, light pollution, etc.

At present, research has focused mainly on the safety and durability of roads. With the development of urbanization and increasing requirements for environmental protection, multiple environmental media, such as water, soil, air, sound, light, and heat in road areas are receiving increasing attention. Improving the sustainability of road pavement has become a bottleneck that needs to be overcome. To comprehensively improve the "multidimensional health" of transportation infrastructure, interdisciplinary research on road ecology is necessary to solve common key scientific problems and core technical bottlenecks. This chapter briefly introduces the development background of sustainable pavement and current sustainable pavement technologies.

8.1 Introduction

8.1.1 Sustainable Development

Sustainable development is an organizing principle for meeting human development goals while also sustaining the ability of natural systems to provide natural resources and ecosystem services on which the economy and society depend.

© Tongji University Press Co., Ltd. 2026

H. Li et al., *Road Engineering*, https://doi.org/10.1007/978-981-95-6659-4_8

In 1983, the World Commission on Environment and Development (WECD) was established, and in 1987, it published "Our Common Future" research report [1], which was also known as the Brundtland Report. The report included a definition of "sustainable development", which is now widely used. Sustainable development is development that meets the needs of the present without compromising the ability of future generations to meet their own needs. Sustainable development, like sustainability, is considered as having three dimensions: the environmental, the economic, and social, as shown in Fig. 8.1. For environmental sustainability, in 2000, the United Nations launched 8 Millennium Development Goals (MDGs) to be achieved by the global community by 2015. Goal seven was to "ensure environmental sustainability", which covered natural resource use, pollution prevention, and biodiversity, etc. For economic sustainability, the general definition of economic sustainability is the ability of an economy to support a defined level of economic production indefinitely, while the defined level of economic production varies among countries. Economic sustainability does not simply refer to gross national product, exchange rates, inflation, and profit; rather, it relates to the production, distribution, and consumption of goods and services. Social sustainability is a critical component of a community's well-being and longevity. Social sustainability is neglected in mainstream sustainability debates. The Oxford Institute for Sustainable Development (OISD) defines social sustainability as follows: Concerning how individuals, communities and societies live with each other and set out to achieve the objectives of development models that they have chosen for themselves, it also takes into account the physical boundaries of their places and the Earth as a whole. The four dimensions of social sustainability are quality of life, equality and diversity, social cohesion, and democracy and governance. The three dimensions are interrelated, interact with each other, and constitute the connotation of sustainable development.

8.1.2 Sustainable Development Goals

The Sustainable Development Goals (SDGs) were adopted in 2015 by the United Nations General Assembly (UN-GA), which are a collection of seventeen interlinked global goals designed to be a "blueprint to achieve a better and more sustainable future for all". They address the global challenges we face, including those related to poverty, inequality, climate change, environmental degradation, peace, and justice. The seventeen goals are all interconnected, and to leave no one behind, it is important that we achieve them all by 2030. The 17 SDGs are as follows: no poverty; zero hunger; good health and well-being; quality education; gender equality; clean water and sanitation; affordable and clean energy; decent work and economic growth; industry, innovation and infrastructure; reduced inequality; sustainable cities and communities; responsible consumption and production; climate action; life below water; life on land; peace, justice and strong institutions; and partnerships for the goals. A diagram listing the 17 Sustainable Development Goals is shown in Fig. 8.2.

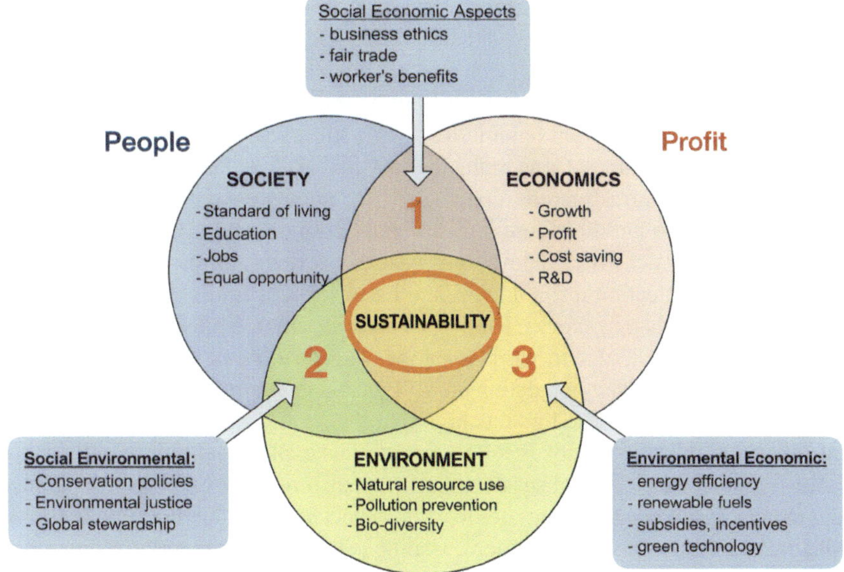

Fig. 8.1 Three dimensions of sustainability

Fig. 8.2 17 UN Sustainable development goals

Various SDGs promote sustainable transport to meet the defined goals. These include SDG 3 on health (improved road safety), SDG 7 on energy, SDG 8 on decent work and economic growth, SDG 9 on resilient infrastructure, SDG 11 on sustainable cities (access to transport and expanded public transport), SDG 12 on responsible consumption and production (ending fossil fuel subsidies) and SDG 14 on oceans, seas and marine resources. Concerning the sustainable development goals report [2], the transportation infrastructure-related goals are mainly covered SDG 9 and SDG 11. SDG 9 aims to build resilient infrastructure, promote inclusive and sustainable industrialization, and foster innovation. A lack of resilient infrastructure, information and communication technologies, and basic services limits a country's ability to perform and adjust to shocks. Rural road connectivity provides farmers and their families with easy access to markets, along with health and education facilities. Vast swaths of the global population are still unable to connect, either through rural roads or cyberspace. These findings highlight the significant role of transportation infrastructure in connectivity. SDG 11 aims to make cities and human settlements inclusive, safe, resilient, and sustainable, which also mentions that poorly planned and managed urbanization translates to a disconnect between the provision of infrastructure and residential concentrations, leading to inadequate networks of streets and a lack of reliable transport systems. As the pandemic response continues, countries and cities need to provide options for accessible, safe, reliable, and sustainable public transport systems.

8.1.3 Sustainable Development of Transportation Infrastructure

The role of transport in sustainable development was first recognized at the 1992 United Nations Earth Summit. Furthermore, at the 2002 World Summit on Sustainable Development, the role of transport was once again captured in the outcome document—the Johannesburg Plan of Implementation (JPOI). The JPOI provided multiple anchor points for sustainable transport in the context of infrastructure, public transport systems, goods delivery networks, affordability, efficiency, and convenience of transportation; improved urban air quality and health and reduced greenhouse gas emissions. Sustainable transport refers to modes of transportation that are sustainable in terms of their social and environmental impacts. Components for evaluating sustainability include the particular vehicles used for road, water, or air transport; the source of energy; and the infrastructure used to accommodate the transport (roads, railways, airways, waterways, canals, and terminals).

The sustainable development of transport infrastructure mainly covers the following aspects: energy consumption, the hydrological cycle, GHG (greenhouse gas) emissions, thermal environmental impact, traffic noise pollution, biodiversity, resource utilization, economic development, etc. Meanwhile, there are a number of research groups working on sustainable transportation worldwide.

(1) The Institute of Transportation Studies at the University of California, Davis (ITS-Davis) leads the National Center for Sustainable Transportation (NCST) in partnership with California State University, Long Beach, the University of California, Riverside, the University of Southern California, the Georgia Institute of Technology, and the University of Vermont. The research initiatives of the NCST are organized in the following four areas (Fig. 8.3): environmentally responsible infrastructure and operations, multimodal travel and sustainable land use, zero-emission vehicle and fuel technologies, and institutional change.

(2) The Transportation Sustainability Research Center at the University of California, Berkeley, was formed in 2006 and has been a leading center in conducting timely research on real-world solutions for more sustainable transportation systems in the future. The TSRC conducts research on a wide array of transportation-related issues, addressing the needs of individuals as well as the public. Research efforts are primarily concentrated in six fundamental areas: advanced vehicles and fuels, energy and infrastructure, the future of mobility, goods movement, mobility for special populations, and shared mobility. The TSRC has assisted in developing and implementing major California and federal regulations and initiatives regarding sustainable transportation.

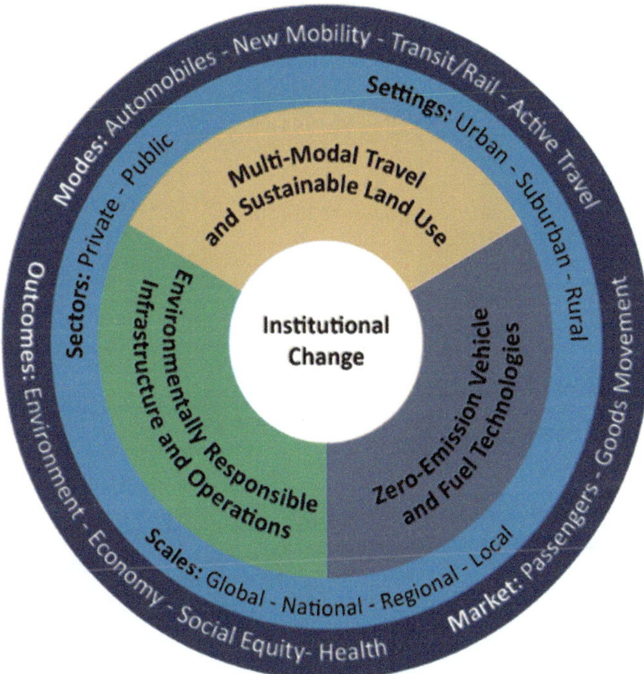

Fig. 8.3 Research initiatives of the national center for sustainable transportation

(3) The China Sustainable Transportation Center (CSTC), founded in 2005, is the technical center for the China Sustainable Cities Program. The CSTC is dedicated to creating sustainable urban and transportation systems, promoting compact land use and transit-oriented development patterns, relieving urban congestion and reducing greenhouse gas emissions, and therefore creating low-carbon, sustainable, and livable cities.

For transportation infrastructure, to make pavements more sustainable, the Federal Highway Administration (FHWA), as a global authority on pavement research, has played a leading role in promoting sustainable transportation and sustainable pavements. In Toward Sustainable Pavement Systems: A Reference Document [3], FHWA proposed that "sustainable" in the context of pavements refers to system characteristics that encompass a pavement's ability to (1) achieve the engineering goals for which it was constructed; (2) preserve and (ideally) restore surrounding ecosystems; (3) use financial, human, and environmental resources economically; and (4) meet basic human needs such as health, safety, equity, employment, comfort, and happiness. Prof. Hui Li also led the research group of the Center for Sustainable Transportation (CST) at Tongji University and completed the translation of this report in 2015. According to the report titled "Built to last: Making sustainability a priority in transport infrastructure" from McKinsey, transportation infrastructure sustainability should be embedded into all stages of the life cycle of transport infrastructure. The pavement life cycle is presented in Fig. 8.4.

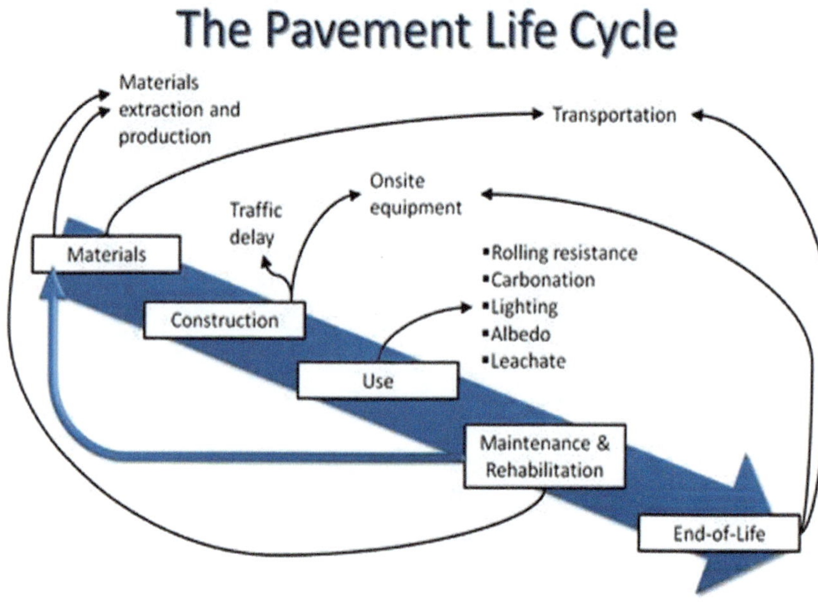

Fig. 8.4 Pavement life cycle

The following presents pavement sustainability solutions for the future in terms of the life cycle.

(1) For the design phase, various design strategies, such as long-life asphalt pavement, long-life cement concrete pavement, open-graded friction course (OGFC), and modular pavement systems, are used to improve pavement sustainability. The design trend includes improving the traditional mechanistic-empirical (M-E) design method, considering the life-cycle economic and environmental impacts, developing new materials, planning future maintenance and reconstruction, integrating performance-related specifications, developing prediction models for improving smoothness, etc.

(2) For the construction phase, the impacts of pavement construction on sustainability include fuel consumption, GHG emissions, noise, traffic delays, and impacts on surrounding residents, businesses, and the local ecological environment. Therefore, the sustainability of pavement construction can be improved in the following ways: optimizing the construction plan and procedure; controlling subgrade erosion and deposition; managing delays, onsite equipment and noise control; managing construction waste; improving the operation efficiency of construction equipment; reducing combustion emissions; and incorporating the Internet of Things. Notably, improving the utilization rate of recycled materials is particularly important. Currently, many pavement projects prioritize recycled materials, such as applying recycled solid waste to each layer of the pavement structure as much as possible and applying recycled asphalt pavement (RAP) to replace both binder and aggregates in new asphalt mixtures rather than just as aggregates. The recycling methods can be further divided into plant-mixed recycling methods and in-place recycling methods.

(3) For the use phase, improving sustainability can be achieved from both intelligent infrastructure materials and road safety. Intelligent facility materials include road condition perception materials, luminous pavement, adaptive lighting systems, and road energy collection, which can significantly reduce the energy consumption of pavements during the use phase. On the other hand, improving road safety can be achieved by adjusting the International Roughness Index (IRI), friction and transverse slope of the pavement. Moreover, adjusting the porosity of pavement materials can also increase road safety. For example, permeable pavement can reduce water accumulation and rain spray, thus preventing vehicles from skidding, hydroplaning, and fog on rainy days to ensure traffic safety.

(4) For the maintenance and rehabilitation phases, it is necessary to combine a pavement management system (PMS) with pavement maintenance. In the case of high traffic volume, the high economic cost caused by frequent maintenance, including road closure and traffic interruption, can be compensated for by reducing the environmental impact of vehicles on the pavement after repair. At low traffic volumes, sustainability can be improved at a lower total life-cycle cost by correct maintenance at the appropriate time. Using life-cycle assessment

(LCA), life-cycle cost analysis (LCCA), and rating systems to assess environmental and social impacts is regarded as an effective method for evaluating the sustainability of different pavement maintenance and rehabilitation strategies.

Sustainable pavement cannot be separated from pavement materials and road structures. The following section briefly reviews the current research on sustainable pavement structures and materials. In addition to sustainability, resilience has become an increasingly important objective for pavement infrastructure. While sustainability emphasizes long-term environmental, economic, and social performance, resilience focuses on the ability of pavement systems to resist, absorb, recover from, and adapt to disruptive events such as heavy rainfall, flooding, heat waves, freeze-thaw cycles, and sudden traffic growth or emergency loading conditions. Therefore, resilient pavement can be regarded as an important extension of sustainable pavement, especially in the context of climate change and increasingly frequent extreme weather events.

8.2   Sustainable Pavement Materials

8.2.1   Application of Industrial Solid Waste in Pavement

Industrial solid waste refers to solid waste with an annual output of more than 10 million tons in the industrial field, which causes serious pollution to the environment or has potential safety hazards, mainly including slag, steel slag, red mud, coal gangue, and fly ash (Fig. 8.5). Moreover, the rapid development of the highway construction industry has increased the consumption of road construction materials such as sand and gravel. If general industrial solid waste is replaced by sand and gravel materials in highway construction, it can not only reduce engineering costs but also promote the utilization of general industrial solid waste, reduce the pressure of solid waste disposal, and reduce environmental pollution and resource waste. However, the performance of industrial solid waste itself is unstable, and there is potential harm to the environment. Therefore, there is an urgent need to establish a scientific application system for industrial solid waste.

Asphalt Rubber Pavement

Waste rubber is the main component of waste tires. It is a kind of polymer elastomeric material that has been difficult to degrade in the natural state for decades, even hundreds of years. Rubber is composed of elastomers (natural rubber and synthetic rubber), vulcanizing agents, vulcanization activators, fillers, carbon black, oil, plasticizers, and additives (antioxidants, antiozonants, etc.). Waste tires have a structure similar to that of rubber, which is a three-dimensional structure formed by a branch chain through crosslinking and polymerization. Rubber has both high elasticity and high viscosity. Processing waste tires into crumb rubber for asphalt modification

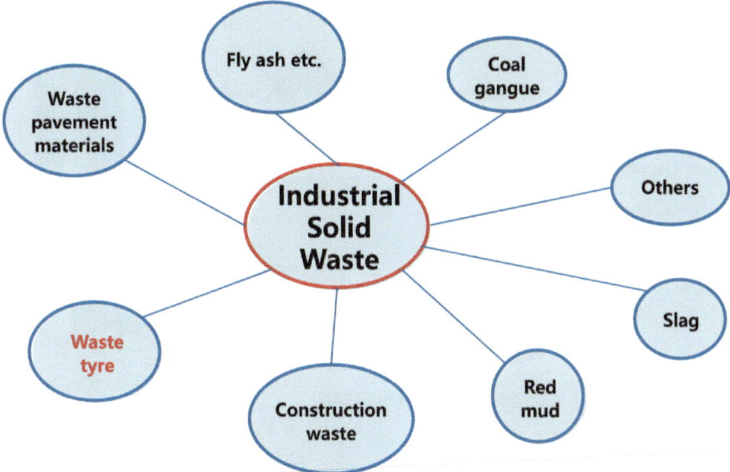

Fig. 8.5 Application of industrial solid waste in pavement

can not only effectively reduce environmental pollution but also reduce the production cost of modified asphalt and endow asphalt pavement with special properties and functions. It is an effective method of resource recycling that integrates energy savings, environmental protection, and new product development [4].

At present, the main modification mechanisms of rubberized asphalt are physical blending and chemical blending. Physical blending refers to the particle effect and interaction effect after the rubber is added to the asphalt [5]. When swelling occurs, a small part of the rubber dissolves and is uniformly dispersed in the asphalt to form a blend system. Chemical blending refers to the desulfurization and cracking reactions of vulcanized rubber in asphalt. The original chemical bonds, such as the sulfur–sulfur bonds in rubber, are broken so that the macromolecular structure is dispersed into a small-molecular chain structure. Under certain conditions, desulfurized rubber molecules react with asphalt molecules to generate new chemical bonds, resulting in chemical crosslinking, thus forming a macromolecular network system [6]. Mixing rubber powder into asphalt to produce rubberized asphalt is the most common application method of rubber powder in the highway industry. Rubber powder from crushed waste tires is added to the pavement material, and wet and dry processes are commonly used. The wet method refers to the process in which rubber powder or particles are mixed with asphalt to produce rubberized asphalt, which is used as a binder and then mixed with mineral materials. The dry method refers to the process of mixing rubber powder with aggregate first, and then spraying it into the asphalt mixture. Rubber powder is directly added to asphalt via the wet method, and the rheological properties of asphalt are improved by the interaction between rubber powder and asphalt. However, specific equipment is needed in the preparation process, and the cost is high. The dry method replaces part of the fine aggregate with rubber powder. Compared with the wet method, the dry process is

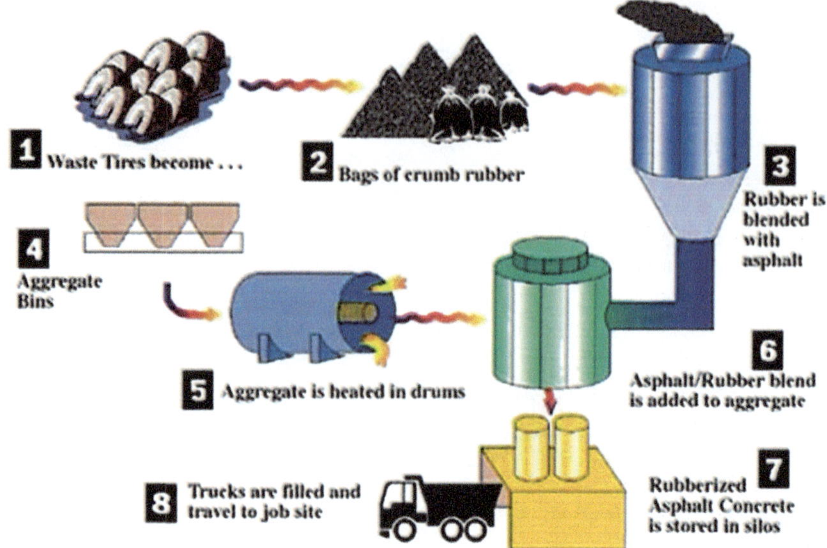

Fig. 8.6   Asphalt rubber pavement material production

simpler, and more rubber powder can be added to the asphalt mixture, which has greater economic benefits. However, owing to the poor bonding strength between rubber powder and asphalt, asphalt mixtures prepared by the dry method need longer curing and performance development times, resulting in the fatigue resistance and stiffness modulus of asphalt mixtures prepared via the dry method containing a large amount of rubber powder being lower than those of conventional asphalt mixtures. Rubber powder ground from waste tires can be combined with other materials to overcome the defects of modified asphalt and enhance the performance of modified asphalt so that the waste can be comprehensively utilized (Figs. 8.6 and 8.7).

Application of Other Solid Waste in Highway Engineering

At present, waste rubber, fly ash, and slag are the most widely studied and technologically mature materials in highway engineering. Air-cooled blast furnace slag is a dense, hard slag formed by the solid waste slag produced in the process of smelting iron in a blast furnace, which is slowly cooled in air or cooled with an appropriate amount of water.

Steel slag is the waste residue discharged in the process of steelmaking. The slag production rate is approximately 15–20% of crude steel output. Porous bricks and permeable bricks made of steel slag have the advantages of light weight, thermal insulation, cool insulation, sound insulation, and extended service life in buildings (Fig. 8.8). Red mud is an industrial waste produced in the refining process of alumina,

Fig. 8.7 Red mud

and its main component is alumina-related residue. Since red mud has compressive properties similar to those of clay and friction properties similar to those of sandy soils, it can be used as a building material and fill material. Red mud is mainly used to replace lime powder and manganese slag admixtures in asphalt mixtures in highway engineering to improve the performance of asphalt mixtures and lay semi-rigid bases [7, 8].

A large number of reclaimed pavement material generated by pavement maintenance are discarded. On the one hand, stockpiled waste pavement materials occupy land resources and pollute the ecological environment. On the other hand, precious asphalt and mineral resources are wasted. If it can be recycled, it will save considerable material costs and bring considerable economic benefits. More importantly, the environmental benefits of asphalt pavement recycling are even greater.

8.2.2 Piezoelectric Pavement

Piezoelectric materials are crystalline materials that can produce a voltage between their two ends when subjected to pressure. Piezoelectric power generation technology uses the direct piezoelectric effect of piezoelectric materials to convert mechanical vibration energy into electrical energy.

Piezoelectric pavement uses piezoelectric energy-harvesting technology to convert part of the mechanical energy generated by vehicle loads into electrical energy. Asphalt concrete pavement must withstand hundreds of thousands to hundreds of millions of repeated axle loads during its design life, resulting in stress,

Fig. 8.8 Permeable brick and concrete made from solid waste

strain, displacement, and a certain degree of vibration. At this time, the pavement will obtain strain energy and kinetic energy from the work of the traffic load and gravity [9]. If a suitable piezoelectric transducer is embedded within the pavement surface, part of the mechanical energy generated by the external force can be converted into electrical energy and further collected and utilized (Fig. 8.9).

8.2.3 Cool Pavement

Cool pavement technology refers to the use of the material's own characteristics to adjust the road-surface temperature by suppressing the temperature of the road surface by means of heat reflection, insulation, absorption, and storage, etc. At present, research on cool pavement technology mainly includes pavement heat-reflection technology, phase-change temperature-regulation technology, thermally resistant pavement, and water-retentive cooling pavement [10, 11] (Fig. 8.10).

Fig. 8.9 Piezoelectric pavement

Fig. 8.10 Cool pavement at Canoga Park, Los Angeles, California

Pavement Thermal Reflection Technology

Pavement thermal reflection technology refers to the use of solar-reflective coatings, thermochromic materials and other methods to change the solar radiation and absorption characteristics of asphalt pavement, thereby reducing the surface and internal temperatures of the road. The basic principle of solar thermal reflection coating is to use the high reflectivity of the constituent materials to reflect the solar radiation to the external space to reduce the solar thermal radiation into the pavement, thereby inhibiting the temperature field of the pavement [12].

Phase-Change Temperature Regulation Technology

Phase-change temperature regulation technology uses phase-change materials to change its phase state so that the road can store heat at high temperatures to realize the purpose of actively regulating the surface temperature of the road. In the past 10 years, many studies have applied phase change materials to asphalt mixtures to obtain temperature-regulating asphalt pavements. Among these materials, organic phase change materials have been widely used in pavement temperature regulation because of their good heat storage performance, non-corrosive properties and nonsupercooling phenomenon [13]. Alcohols, alkanes and paraffins are commonly used as the main phase change materials for temperature-regulated pavements, and they are usually assisted by shaping and encapsulation technology to reduce the leakage of phase change materials during use [14].

Other Pavement Temperature Regulation Technologies

The aim of thermally resistant pavements is to reduce the thermal conductivity of pavement aggregates, prevent heat transfer from the pavement surface to the interior of the pavement, and reduce the pavement temperature. The commonly used thermally resistant aggregates include ceramsite, vermiculite, bauxite and so on. The thermal resistance of asphalt pavement designed by directly replacing ordinary gravel can effectively reduce the temperature in the pavement structure, but the pavement performance can be reduced to varying degrees, especially poor water stability.

8.2.4 Self-illuminating Pavement

Road lighting aims to provide drivers and pedestrians with safety and traffic guidance, but the large-scale application of road lighting facilities has led to increasingly high energy consumption and light pollution. The glare caused by traditional street lighting can cause visual discomfort and fatigue in drivers and passengers and can even cause traffic safety hazards in severe cases. The use of embedded light-emitting devices on the road or doping self-luminous materials to make the road emit light autonomously, replacing all or part of the peripheral lighting sources, can not only save energy but also eliminate the uneven illumination and glare caused by streetlamps.

Researchers have achieved the goal of lighting concrete roads at night by adding phosphors to the concrete, changing the microstructure of the concrete, and coating the surface of the concrete with luminescent materials [15]. This new type of functional pavement with luminous materials is safe, beautiful and energy-saving. At present, fluorescent powder, artificial luminescent stones, strontium aluminate, luminous paints, luminous glazes, and other luminous materials are used in pavement. Considering the performance requirements and construction technology of pavement, fluorescent powders and artificial luminescent stones are the most widely used. As a smart functional pavement, self-illuminated pavement can effectively improve the visibility of night pavement without increasing energy consumption.

At present, research on self-illuminating pavement worldwide is still in its infancy, and many technical and application problems still need to be solved. For example, owing to the special properties of phosphor powder, the use of an asphalt mixture mixed with phosphor powder cannot achieve luminescence. At present, inorganic cement materials are mostly used as light-emitting carriers of phosphor powder. The range of applicable materials is relatively limited, and the method causes high driving noise and cannot guarantee the high-temperature stability and low-temperature crack resistance of the pavement material. It is necessary to focus on how to improve the performance of luminescent materials to achieve as much performance as possible. Therefore, the scientific design of self-luminous pavement materials, the selection of self-luminous pavement structures, and the long-term evaluation mechanism of the luminous performance of self-luminous pavements need to be further studied [16] (Fig. 8.11).

Fig. 8.11 Self-luminous pavement at Jinhua, Zhejiang

8.2.5 Snowmelt Pavement

Active snow-melting pavement refers to changing the material composition and structural design of traditional pavement so that the pavement can actively complete snow melting without the help of external effects in the snowfall process. This approach can not only avoid reducing the efficiency of traffic operations and frost heave and mud pumping, but also improve traffic safety. At present, research on snowmelt pavement has focused mainly on self-stressing elastic roads, energy-conversion roads, pavement coatings with freezing-point-depressing additives, and active deicing and snow pavement coating technologies based on superhydrophobic materials [17, 18].

The self-stressing elastic pavement changes the contact characteristics between the pavement and the tire by adding a certain amount of highly elastic material to the pavement. Through the high deformation characteristics of elastic materials, the stress generated by traffic loads on the road surface removes snow and breaks accumulates ice. This technology includes rubber-particles asphalt pavement and mosaic pavement technology. In addition, increasing the depth and roughness of the pavement structure can make it difficult for the ice and snow layers to freeze under the uneven stress of vehicle loads. For example, the application of porous asphalt concrete roads can effectively keep pavement clean. Energy-conversion for deicing pavement involves laying heating pipes and cables on it and increasing the pavement temperature through heat generated by electricity, solar panels or natural gas heating technology to achieve the purpose of melting road ice and snow or preventing road freezing. In addition to the above two active deicing technologies, the purpose of melting ice and snow can also be achieved by adding low-freezing-point materials to pavement materials (Fig. 8.12).

8.2.6 Tailpipe Emission Degradation Pavement

Pavement tailpipe-emission-degradation technology uses reusable catalytic materials on the pavement surface to reduce the influence of harmful gases, oil pollution, heavy metals and other pollutants and alleviate the problem of environmental pollution in the pavement. The most widely used technology is the photocatalytic method [19]. TiO_2 is one of the most widely used photocatalytic materials because of its advantages, such as spectral response and catalytic efficiency. By applying photocatalytic materials to pavement, harmful gases such as CO, HC and NOx in the road surface environment are catalyzed and degraded under the action of sunlight and then oxidized into carbonates and nitrates, which are absorbed by pavement and washed by rainwater or artificial flushing to purify the gas environment [20].

Nono-TiO_2 is sprayed on pavement, and the photocatalytic properties of nanotitanium dioxide are used to form the tailpipe-emission-degradation pavement. To avoid the influence of TiO_2 on the pavement performance of asphalt pavement, TiO_2 -based photocatalytic decomposition via automobile exhaust technology is combined

Fig. 8.12 Snowmelt pavement at Daxing airport, Beijing

with thin pavement overlays and microsurfacing technology, and a thin cover layer
with a thickness of 5–25 mm is formed on the original pavement, which can achieve
exhaust gas degradation and improve road performance (Fig. 8.13).

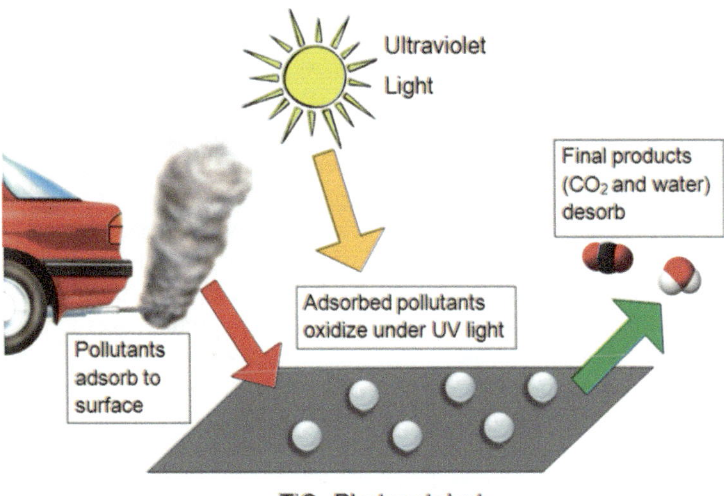

Fig. 8.13 TiO$_2$ photocatalyst for the degradation of tailpipe emissions from pavement

8.2.7 Runoff Pollution Purification Pavement

The technology of permeable asphalt pavement aims mainly to purify and reduce pollutants such as phosphorus, sulfur, and heavy metals in road runoff pollution and prevent groundwater pollution by adding materials with purification functions to the pavement of permeable pavement structure [21]. At present, the purification layer materials used in permeable asphalt pavement mainly include zeolite, activated carbon, ceramsite and fine sand [22].

The typical runoff pollutant removal mechanism mainly includes physical interception, adsorption, and biological action [23]. These include (1) the retention of particulate organic matter by the structural layer; (2) the adsorption of organic matter by the granular medium; (3) the adsorption of organic pollutants by biological flocs on the surface of the medium particles; and (4) the oxidative degradation of microorganisms. The residence time of infiltrating rainwater in permeable pavement is very short, and the biological effect can be ignored. The removal of road runoff pollutants is based on physical interception and adsorption. During the purification process of road runoff and infiltration rainwater, the pores in paved roads trap suspended solids in the runoff, and the runoff pollutants are removed to a certain extent. Soluble pollutants in the runoff are adsorbed by particulate matter and then intercepted and retained inside the paved road when they infiltrate through the pores of the permeable pavement (Fig. 8.14).

8.2.8 Photovoltaic Pavement

Pavements cover most of the ground surface, and asphalt roads can directly absorb solar radiation. When photovoltaic power generation technology is used to assist in road construction, it can effectively relieve the pressure on the energy supply without increasing the amount of land. The top layer of photovoltaic pavement is a new translucent material like frosted glass, which is a new pavement structure with the functions of energy harvesting and conversion.

At present, the commonly used photovoltaic pavement structures are the solid-panel structure and the hollow-panel structure. In terms of the solid plate structure, scholars have focused on the structural strength of photovoltaic panels; the skid resistance, flatness and fatigue resistance of light-transmitting materials; and the bonding and drainage performance of the bottom connecting layer to improve its durability and road performance when it is subjected to vehicle loading. Further research on the durability of the structure, material and power-collection circuit of the photovoltaic pavement solid plate is still needed. Laying photovoltaic panels on the road may affect the road performance. How to ensure the rational application of photovoltaic power generation technology while ensuring the macroscopic stability of the overall structure of the road will become the focus of future research (Fig. 8.15).

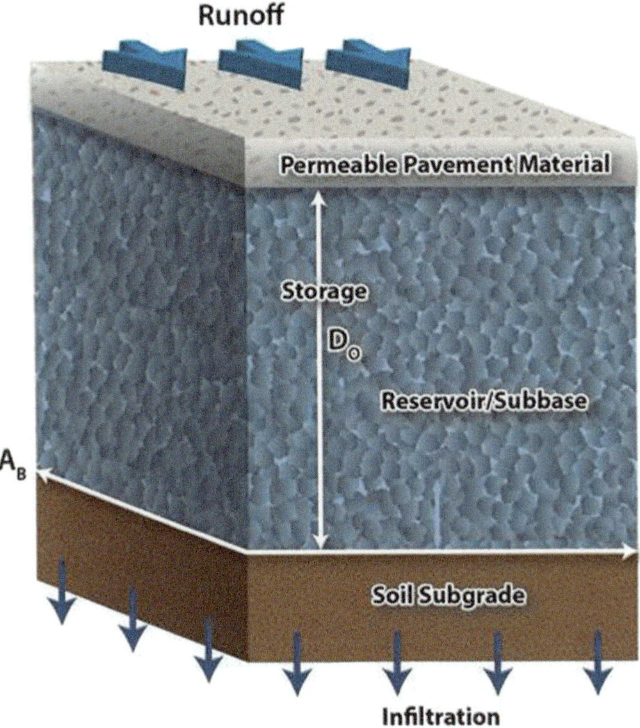

Fig. 8.14 Runoff pollution purification pavement

Fig. 8.15 The top layer of photovoltaic pavement

8.2.9 *Nanomaterial Pavement*

For traditional asphalt pavement materials, polymer modifiers are usually added to improve the performance of asphalt materials, such as waste rubber powder, SBS, SBR, etc. Although they can effectively improve the performance of bituminous materials, they also cause a decrease in the aging resistance and poor storage stability of the mixture, which limits the degree of modification of bituminous materials. With increasing understanding of nanomaterials in recent years, an increasing number of researchers have begun to pay attention to the application of nanomaterials in pavement.

Nanomaterials are materials whose structural units have at least one dimension in the range of 1–100 nm. In addition, according to their varied sizes, nanomaterials can be divided into zero-dimensional, one-dimensional, and two-dimensional nanomaterials. Owing to their size, nanomaterials have many excellent properties not found in traditional materials, such as the macroscopic quantum tunneling effect, large specific surface area, high surface free energy, good dispersion and so on. More importantly, nanomaterials have strong anti-aging properties, self-cleaning properties, and ability to purify vehicle exhaust emissions. These effects increase the efficiency of modifying asphalt to meet the demands of transportation development. Therefore, researchers have added nanomaterials as asphalt modifiers to improve the high-temperature performance, low-temperature performance, water sensitivity, and aging resistance of asphalt. At present, nano-modified asphalt that has been widely used in pavement can be divided into two categories: pure nano-modified asphalt and nanomaterial composite-modified asphalt [24] (Fig. 8.16).

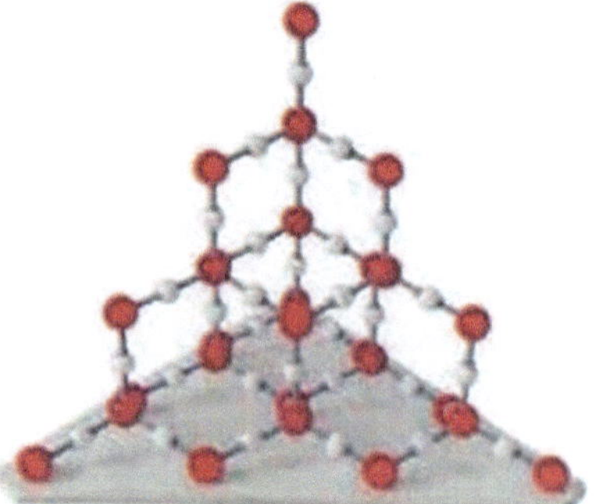

Fig. 8.16 Nanosilica

Nano-modified Asphalt

The pure nano-modified asphalt modifiers include nano-$CaCO_3$, nano-TiO_2, nano-Fe_3O_4, and nano-layered silicate. These modifiers are improved by directly adding a certain amount of nano-modifier to the base asphalt. The basic properties of base asphalt improve the performance of asphalt pavement.

Nano-composite-Modified Asphalt

Nano-composite-modified asphalt relies on nanomaterials and polymer modifiers to improve the basic technical indicators of matrix asphalt through physical or chemical changes to improve the comprehensive performance of asphalt pavement. Common nanomaterial composite modifiers include nano-ZnO/SBS composite modifiers, nano-layered silicate/SBS composite modifiers, and carbon nanotube/SBS composite modifiers.

Other Applications of Nanomodified Asphalt

With the advancements in industrialization and environmental problems in China, researchers have reported that the photocatalytic effect and self-cleaning properties of nano-TiO_2 can improve the air quality. Relevant experimental studies have shown that the incorporation of nano-TiO_2 not only significantly improves the light- ageing resistance of asphalt but also has a good tailpipe-emission degradation effect on OGFC asphalt mixtures with a nano-TiO_2 content of 60% (proportion of mineral powder) and microsurfacing mixtures with a nano-TiO_2 content of 50%. In addition, the degradation effect of the nano-TiO_2 material on automobile exhaust showed that the incorporation of a certain proportion of nano-TiO_2 under light could react with the carbon oxides and nitrogen oxides in automobile exhaust so that the nitrogen oxides in automobile exhaust could be effectively degraded.

8.3 Sustainable Pavement Structures

8.3.1 3D Painted Pavement

3D pained pavement simulates three-dimensional visual effects on a two-dimensional plane. The 3D artistic pavement is different from traditional color (anti-skid) pavement products because of its two-dimensional plane composition, giving the viewer a subversive three-dimensional visual impact. Three-dimensional pained pavement coatings are pigments that are made of high-molecular-weight polymer materials with high color saturation and strong weather resistance and are suitable for outdoor

Fig. 8.17 3D cross walk

painting. Moreover, good skid resistance properties also make it compatible with the ground. The medium is highly fit. The creative method of 3D painting not only strengthens the warning function of road signs but also plays a better role in beautifying the city. For example, 3D painted crossings can allow drivers to notice them from farther away, thereby reducing driving speed and reducing accidents (Fig. 8.17).

8.3.2 Bicycle Pavement

The bicycle transportation system usually consists of bicycle lanes and supporting parking facilities. It is a low-carbon, environmentally friendly and healthy transportation mode, and it is also an important part of slow traffic systems. It can not only effectively alleviate the urban surface traffic burden and reduce air pollution and energy consumption but also enhance residents' sense of travel pleasure and happiness, which plays an important role in the sustainable development of cities [25].

Bicycle pavements have high requirements for surface smoothness and landscape quality (Fig. 8.13). First, bicycle pavement has sufficient strength, stability, smoothness, and skid resistance to meet the driving requirements of bicycles. Second, certain requirements for visual guidance and segregated traffic operation should be met to satisfy the function of bicycle lane ownership. Finally, bicycle pavement

Fig. 8.18 Bicycle pavement

should apply low-carbon and environmentally friendly technologies to meet the requirements of users (Fig. 8.18).

8.3.3 Smart Pavement

Smart pavement is a modern roadway system that integrates the development of informatization and intelligence throughout the entire chain of transportation planning, construction, operation, service, and supervision. Networking, mobile internet, intelligent control and other technologies are deeply integrated with transportation to improve highway service levels, reduce environmental pollution, improve road construction and operation safety, and improve driving comfort and safety (Fig. 8.19). From the perspective of resilience, smart pavement can also improve the ability of road infrastructure to monitor hazards, provide early warning, support rapid decision-making, and enhance post-event recovery under extreme environmental and traffic conditions.

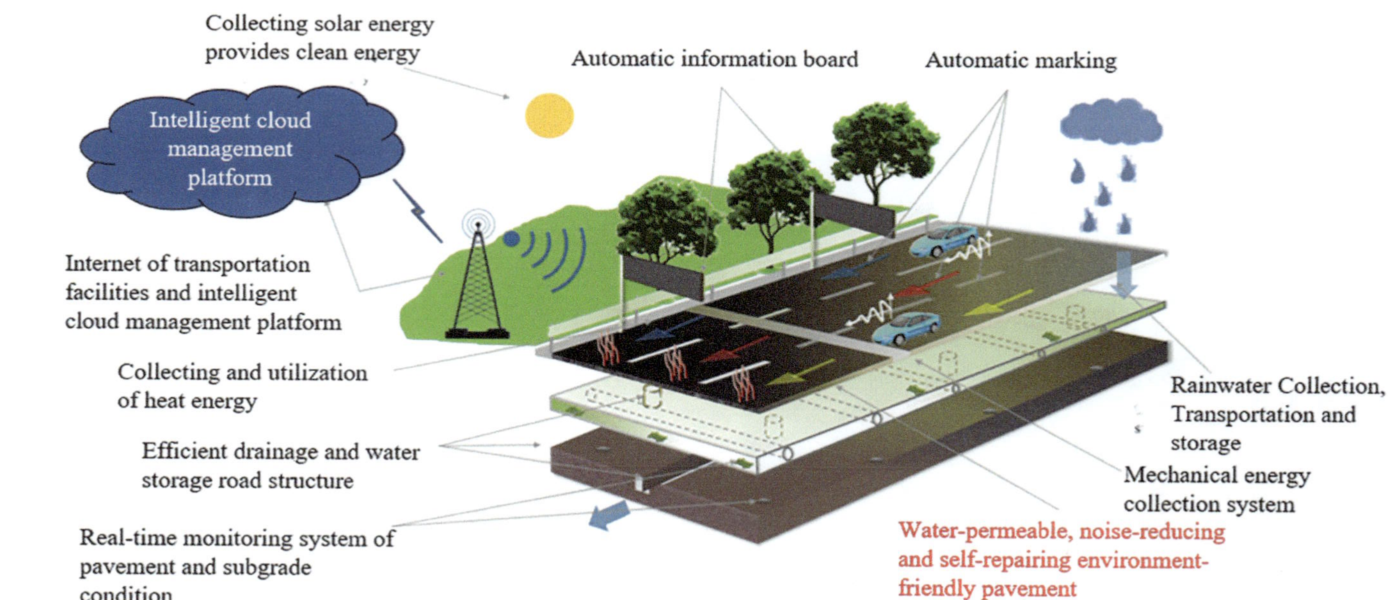

Fig. 8.19 Smart road

Self-Healing Pavement Materials

Self-healing materials can be divided into two categories according to their properties: asphalt-based self-healing materials and cement-based self-healing materials. Owing to its own rheological properties, asphalt can repair itself after a certain amount of damage, but asphalt flows slowly at room temperature and has limited self-healing ability. Scholars in China and abroad have carried out various studies on improving the self-healing ability of asphalt. At present, mainstream asphalt self-healing technologies include the microcapsule method and electromagnetic heating method [26]. Cement concrete pavement can be divided into passive repair and active repair methods according to different self-healing methods. Passive repair technology includes engineering repair, cement-based composite material repair, mineral admixture repair and microbial repair. In terms of active repair, shape-memory alloys are currently used more intelligently in the field of civil engineering.

Self-Monitoring Pavement Materials

Real-time understanding of road damage is extremely important in the design, construction, and operation of highways. The road safety monitoring method based on manual periodic inspection is time-consuming and labor-intensive, affects road traffic, and cannot obtain road conditions in real time. Self-monitoring intelligent pavement materials not only possess perceptual driving characteristics but also have more advanced characteristics, such as self-healing, information processing and self-adaptation.

Other Multifunctional Pavement Materials

In addition to self-healing and self-monitoring pavement materials, a number of multifunctional pavements have emerged in recent years to meet specific needs. When energy is increasingly scarce and the world is vigorously advocating energy conservation and emission reduction, photovoltaic pavement has naturally become a research hotspot in the field of pavement materials. Owing to the functional requirements of special sections of highway asphalt pavement, the research and development of energy-harvesting pavements, self-melted pavements, self-healing pavements, and other related methods will be the research hotspots of multifunctional pavement materials [27].

8.4 Future Directions in Sustainable Pavement Development

The evolution of sustainable pavement systems is poised to redefine transportation infrastructure in the coming decades. Building on the advancements outlined in the previous sections—ranging from recycled materials to smart pavements, the future of sustainable pavement development will hinge on addressing emerging challenges such as climate resilience, resource efficiency, and technological integration. This section explores key innovations and strategic priorities that will shape the next generation of sustainable roadways.

The ongoing shift toward sustainable pavement materials will continue to be one of the most significant trends in road development. Traditional materials such as asphalt and concrete have been widely used because of their durability and cost-effectiveness. However, they are also major contributors to carbon emissions and environmental degradation. In the future, pavements will increasingly rely on eco-friendly materials, such as recycled aggregates, bio-based binders, and self-healing asphalt. Innovations such as cold-mix asphalt technologies, which eliminate the energy-intensive heating of traditional asphalt, are gaining momentum. Similarly, carbon capture and utilization (CCU) techniques—where CO_2 is injected into concrete or asphalt during production—are being tested to create low-carbon or potentially carbon-negative pavements.

Technological advancements in digitalization will play a transformative role in pavement infrastructure. Smart roads, which are equipped with sensors and communication systems, enable real-time monitoring of traffic conditions, environmental factors, and pavement health. These roads are integral to the development of connected and autonomous vehicles (CAVs), as they facilitate seamless communication between vehicles and infrastructure to improve traffic flow and reduce accidents. These intelligent systems can also provide insights into pavement maintenance needs, reduce costs and improve the lifespan of the pavement infrastructure. The integration of artificial intelligence (AI) and machine learning into pavement management systems (PMs) represents a paradigm shift in infrastructure. By analyzing data from embedded sensors, drones, and satellite imagery, AI algorithms can predict pavement deterioration with unprecedented accuracy.

As climate change continues to influence weather patterns, climate-resilient pavement design will become increasingly important. In addition to climate adaptation, future pavement systems should also emphasize rapid recovery, service continuity, and adaptive maintenance after disruptive events, so that resilience can be incorporated into the full life cycle of sustainable pavement. The development of pavements that can withstand extreme weather events, such as heavy rainfall, flooding, and heat waves, will be crucial for ensuring the longevity and safety of infrastructure. Climate-responsive materials and adaptive construction techniques will help mitigate the impacts of these conditions. Moreover, roads need to be designed with greater emphasis on stormwater management and flood control, particularly in regions prone to flooding or drought. Future pavement design will incorporate green infrastructure solutions such as permeable pavements to reduce the urban heat island effect and enhance environmental sustainability.

The road ahead for sustainable pavement systems is both challenging and transformative. By merging technological ingenuity with ecological stewardship, the industry can deliver infrastructure that not only supports economic growth but also regenerates ecosystems and enhances societal well-being. The development of these advanced solutions requires not only technological advancements but also policy support, industry collaboration, and a commitment to creating sustainable, future-ready road networks that support both people and the planet.

Discussion, Questions and Exercises

(1) Describe measures to improve the sustainability of pavement from a life cycle perspective. In addition to sustainable pavement structures and materials, what other solutions can be adopted in your opinion?

(2) Explain the methods of adding the rubber powder from the crushed tires to the pavement material and the differences between them.

(3) Describe the negative effects of phosphor powder as a luminescent material on pavement performance. In addition, what other disadvantages of phosphor powder can you think of?

(4) Elaborate the role of permeable asphalt pavement in the purification process of road runoff rainwater and infiltration rainwater.

References

1. BRUNDTLAND G H. Our Common Future—Call for Action [J]. Environmental Conservation, 1987, 14(4): 291–4
2. The Sustainable Development Goals Report 2022 [R]. United Nations, 2022
3. VAN DAM T J, HARVEY J, MUENCH S T, et al. Towards Sustainable Pavement Systems: a Reference Document [R]: United States. Federal Highway Administration, 2015
4. YU H Y, MA T, WANG D W, et al. Review on China's Pavement Engineering Research· 2020 [J]. China Journal of Highway and Transport, 2020, 33(10): 1–66
5. PRESTI D L. Recycled Tyre Rubber Modified Bitumens for Road Asphalt Mixtures: A Literature Review [J]. Construction and Building Materials, 2013, 49: 863–81
6. LI T G, LI J Z, LI W. Micro-mechanism Study and Road Engineering Application of Rubber Asphalt [J]. Journal of Highway and Transportation Research and Development, 2011, 28(1): 25–30
7. ZHANG H, LI H, ABDELHADY A, et al. Optimum Filler–bitumen Ratio of Asphalt Mortar Considering Self-healing Property [J]. Journal of Materials in Civil Engineering, 2019, 31(8): 04019166
8. ZHANG H, LI H, ZHANG Y, et al. Performance Enhancement of Porous Asphalt Pavement using Red Mud as Alternative Filler [J]. Construction and building materials, 2018, 160: 707–13
9. ZHAO H Z, LIANG Y H, LING J M. Study on Harvesting Energy from Pavement Based on Piezoelectric Effects [J]. Journal of Shanghai Jiaotong University, 2011, (S1): 62–6
10. XIE N, LI H, ABDELHADY A, et al. Laboratorial Investigation on Optical and Thermal Properties of Cool Pavement Nano-coatings for Urban Heat Island Mitigation [J]. Building and Environment, 2019, 147: 231–40

11. LI H. Evaluation of Cool Pavement Strategies for Heat Island Mitigation [M]. University of California, Davis, 2012

12. ZHENG M L, HE L T, GAO X et al. Analysis of Heat-reflective Coating Property for Asphalt Pavemnet Based on Cooling Function [J]. Journal of Traffic and Transportation Engineering, 2013

13. TAN Y Q, BIAN X, SHAN L Y, et al. Preparation of Latent Heat Materials Used in Asphalt Pavement and Its Performance of Temperatire Control [J]. Journal of Building Materials, 2013, (2): 354–9

14. HE L H. Preparation and Cooling Mechanism of Composite Phase Change Heat Storage Asphalt Pavement Materials [D]. Chongqing: Chongqing Jiaotong University, 2016

15. SAHU I P, BISEN D, BRAHME N, et al. Luminescence Studies of Dysprosium Doped Strontium Aluminate White Light Emitting Phosphor by Combustion Route [J]. Journal of Materials Science: Materials in Electronics, 2015, 26(11): 8824–39

16. SHA A M, JIANG W, WANG W T, et al. Design and Prospect of New Pavement Materials for Smart Road [J]. Chinese Science Bulletin, 2020, 65: 3259–69

17. ZHANG H W, HAN S, LIU H H. A Summary of Asphalt Concrete Pavement for Deicing and Snow Melting Technology [J]. Communications Science and Technology Heilongjiang, 2008, 31(3): 8–9

18. TAN Y Q, ZHANG C, XU H N, et al. Snow Melting and Deicing Characteristics and Pavement Performance of Active Deicing and Snow Melting Pavement [J]. China Journal of Highway and Transport, 2019, 32(4): 1

19. PEI J Z, WANG Y S, ZHU C D, et al. Research Progress on Automobile Exhaust Pavement Purification Materials [J]. China Journal of Highway and Transport, 2019, 32(4): 92–104

20. GUAN Q, CHEN M. Study on Decontamination of Vehicle Emissions by Nano-TiO_2 Sprayed on Highway Concrete Pavement [J]. Journal of Highway and Transportation Research and Development, 2009, 26(3): 154–8

21. SONG Q, XU Y. Purification Effectiveness of Runoff Pollution by the Permeable Asphalt Pavement [J]. Journal of Northeast Agricultural University, 2009, 40: 56–9

22. KANG A, MAO H, LI B, et al. Investigation of Selective Filtration Characteristics of Filter Media for Pavement Runoff Treatment [J]. Journal of Cleaner Production, 2019, 235: 590–602

23. LI H, LIU J, ZHANG H, et al. Investigation on the Effect of Fine Solid Wastes on the Runoff Purification Performance of Porous Asphalt Mixture [J]. Journal of Environmental Management, 2021, 300: 113612

24. YANG J, TIGHE S. A Review of Advances of Nanotechnology in Asphalt Mixtures [J]. Procedia-Social and Behavioral Sciences, 2013, 96: 1269–76

25. AGARWAL A, ZIEMKE D, NAGEL K. Bicycle Superhighway: An Environmentally Sustainable Policy for Urban Transport [J]. Transportation Research Part A: Policy and Practice, 2020, 137: 519–40

26. SUN D, SUN G, ZHU X, et al. A Comprehensive Review on Self-healing of Asphalt Materials: Mechanism, Model, Characterization and Enhancement [J]. Advances in colloid and interface science, 2018, 256: 65–93

27. WU J Q, SONG X G. Review on Smart Highways Critical Technology [J]. Journal of Shandong University(Engineering Science), 2020, 50(4): 52–69

<u>GPSR Compliance</u>

The European Union's (EU) General Product Safety Regulation (GPSR) is a set of rules that requires consumer products to be safe and our obligations to ensure this.

If you have any concerns about our products, you can contact us on ProductSafety@springernature.com

In case Publisher is established outside the EU, the EU authorized representative is:

Springer Nature Customer Service Center GmbH
Europaplatz 3
69115 Heidelberg, Germany

Batch number: 10256381

Printed by Printforce, the Netherlands